Omega-3 for Optimal Life:

Why You Need FISH OIL

Joseph C. Maroon, MD, FACS and
Jeff Bost, PAC

ISBN: 978-1-4834-5521-1 (sc)
ISBN: 978-1-4834-5522-8 (e)

Library of Congress Control Number: 2016911737

Lulu Publishing Services rev. date: 10/04/2016

Contents

About the Authors

Joseph C. Maroon, MD, FACS is Clinical Professor and Vice Chairman of the Department of Neurological Surgery and Heindl Scholar in Neuroscience at the University of Pittsburgh Medical Center. In addition to being a renowned neurosurgeon, he is a sports medicine expert, health and nutrition advisor and Ironman triathlete. Dr. Maroon has done extensive research in the use of omega-3 fatty acids and co-authored the book, *Fish Oil: The Natural Anti-Inflammatory*, in 2006. This book highlights the many benefits of fish oil known at that time. His other interests include the dietary supplement resveratrol, found in red grape skins, which has been shown to activate genes for improved health. He authored the book, *The Longevity Factor: How Resveratrol and Red Wine Activate Genes for a Longer and Healthier Life* in 2009. In 2012, this book was featured in a PBS special, *The Secrets of Longevity*, shown on TV stations throughout the country.

Dr. Maroon is the author of six books, and co-author of forty book chapters and over two hundred and ninety published scientific papers. He has given more than one hundred and fifty presentations at national and international conferences and is often invited to be a visiting professor at major universities and keynote speaker.

Dr. Maroon has successfully maintained his personal athletic interests through participation in nine marathons and more than seventy-two Olympic-distance triathlon events. However, his greatest athletic accomplishment is his participation in eight Ironman triathlons. Through his intensive involvement with athletics, both as a triathlete and as a consultant to numerous sports teams, Dr. Maroon has a personal interest in healthy living and healthy nutrition. He daily "practices what he preaches."

In 1989, Dr. Maroon (along with NFL great Joe Montana and NBA legend Kareem Abdul-Jabbar) was inducted into the Lou Holtz Upper Ohio Valley Hall of Fame for his athletic accomplishments and contributions to sports medicine. Eleven years later in 2010 he was inducted into the National Fitness Hall of Fame in Chicago. Other inductees include Gov. Arnold Schwarzenegger, Jack LaLanne, and Kenneth Cooper, founder of the Aerobic Movement. Of his accomplishments, in 2011 Dr. Maroon was selected as a "Distinguished Alumnus" of Indiana University—one of five selected annually from five hundred thousand alumni of the University.

Jeff Bost, PAC, is a Neurosurgical Physician Assistant and Clinical Instructor in the Department of Neurosurgery at University of Pittsburgh Medical Center and Clinical Assistant Professor at Chatham University, Pittsburgh, PA. He has been working directly with Dr. Joseph Maroon as an associate since 1987. Dr. Maroon and Mr. Bost have embraced the use of natural supplements in their neurosurgical practice and have written extensively on omega-3 and resveratrol for pain relief and for anti-aging. Mr. Bost has presented sixty-five invited lectures, fifty-five national posters, twenty-nine coordinated research projects, five workshops, and has written thirty-five refereed articles and twenty-four book chapters. He also co-wrote two books on the use of natural supplements, *Fish Oil: The Natural Anti-Inflammatory,* Basic Health Publications 2006 and *The Science of Cocoa for a Healthier You: Scientific Breakthroughs about the Health Benefits of Cocoa and Chocolate,* self-published 2011.

Introduction

As a practicing neurosurgeon, I typically treat patients who have experienced a disease or injury to their brain, spine or nerves. Thanks to modern science we now have many sophisticated procedures and medications to help improve the quality of life for these patients. Despite the vicissitudes endemic in the practice of medicine today, I remain proud to be part of a profession that continually searches for new and better options for healing.

Although I was able to witness and participate in many of these modern *medical miracles*, I became increasingly frustrated that healthcare too often is focused on *fixing things after they are broken* and was doing very little to prevent disease and injury. Unfortunately, our healthcare system is less about *healthcare* and more about *"sickcare."*

These frustrations eventually led me to explore the benefits of alternative treatments that are focused on disease prevention and wellness. Much to my surprise and delight I found thousands of scientific articles and convincing research proving that numerous treatments, many used for thousands of years, could help people stay well and avoid disability and illness.

What I repeatedly discovered in my research were reports on the remarkable benefits of omega-3 fatty acids—and the consequences if deficient. Could fish oil be a key factor to improve health and prevent disease in people? Commonly found in abundance in fish, the omega-3 molecule is an essential structural part of every living cell in our body. We cannot function at our best without adequate levels of omega-3 fatty acids. Despite this fact, the majority of Americans are deficient in this essential oil.

As I explored deeper, I found fish oil could particularly benefit many of the chronic health conditions we develop. Some of the costliest healthcare problems—heart disease, joint degeneration, intestinal problems and even memory problems—may benefit from fish oil and omega-3s. Fish oil has been shown to help all groups of people, women, men, young, old, new mothers, and even babies before they are born.

Fish oil's universal actions and benefits make it one of the best choices for all Americans to influence their health for the better. Taking fish oil as a supplement or as part of a healthy diet, along with adequate amounts of exercise and stress reduction, can save lives, prevent disease, and reduce our healthcare requirements.

In this book, we will discuss the latest science on the benefits of omega-3s and how this special molecule can benefit every organ in our body. We will also focus on how our body's innate natural protective mechanism, the inflammatory response, is hijacked by our poor diet and lifestyle choices and contributes, rather than prevents diseases of aging. We will give you the information you need to make better choices when looking for fish oil supplements and help you determine what dose of omega-3s works best for you. We hope to lead you down a road of exciting discovery, give you the tools to make better choices and help you to die young… as late as possible!

Joseph C. Maroon, MD

CHAPTER

1

History of Fish Oil:
Heart Health Benefits

Let's start at the beginning. Although the scientific history of the benefits of fish oil and omega-3 are recent, there is evidence that from the beginning of time the omega-3 molecule has played a critical role in the development of life on earth. This is true for both plants and animals since omega-3 molecules are found within the cell membranes of almost every life form on earth. Because the scope of learning about omega-3 molecules can be overwhelming, this chapter will provide you with the facts most important to the human health related benefits of omega-3 starting with the heart and vascular system.

Originally, the benefits of omega-3 molecules were discovered in the early 1970s in native peoples, such as the Inuit Eskimos of Greenland, whose diet consisted mostly of seals, whale, and fish. This population was found to have a very low incidence of heart disease and stroke yet their diet often consisted of raw fat, which until only recently was believed to be associated with an increase in heart disease. Contrary to what was preached by many authorities for decades, "good" fats as found in fish oil, walnuts and avocadoes are cardio-protective and carbohydrates (sugars) are the culprit. This ironic discovery resulted in thousands of scientific articles since then demonstrating an expansive array of health benefits from omega-3 essential fatty acids (EFAs). These studies have confirmed that omega-3 EFAs can indeed benefit healthy hearts to prevent disease, but it can also aid those with a higher risk of vascular disease.

Amazing Reported Benefits of Omega-3 EFA (Fish Oil)

- Lowers blood pressure
- Reduces triglycerides
- Slows vascular plaque development
- Reduces incidence of arrhythmia, heart attack, sudden death and stroke
- Reduces postpartum and other forms of depression
- Improves memory
- Reduces vision loss, dementia, and Alzheimer's disease risk
- Reduces joint pain and inflammatory response in autoimmune diseases

Background: Knowing "Good" Fats from "Bad" Fats

In order to understand the health benefits of omega-3 essential fatty acid there are some basic facts you must know. First, not all fats we eat are the same. They can be generally divided into "bad" fats and "good" fats. There are two main types of potentially harmful dietary fats: saturated and artificial manmade fats that contain trans fatty acids. Once fats enter the body they are called lipids and are generally carried in our blood by a protein called a lipoprotein. Cholesterol is an example of a lipoprotein.

Saturated fats are mainly from animal sources of food, such as red meat, poultry and full-fat dairy products. These are the fats that remain solid even at room temperature. Feedlot beef compared to grass-fed generally has a greater percentage of this type of fat. Excessive consumption of saturated fat can result in increased cholesterol levels, especially the LDL (low-density lipoprotein) type of cholesterol. The link between LDL and cardiovascular disease has led to the rapid use of statin drugs designed to lower these "bad" fats. In fact the current value of the cholesterol-lowering drug industry is $29 billion dollars. Ironically, despite the lowering of cholesterol the number of people in the sixty-five to seventy-four age group have about the same rate of heart disease. Some physicians now consider the link between elevated cholesterol and heart disease weak. It is a reminder of how a high-carbohydrate diet was recommended for decades as the most "healthy" diet!

Trans fats, also called partially hydrogenated fats, are manmade from oils and are less likely to spoil or become rancid than naturally occurring oils.

They have been used now for decades in many common processed foods. It has only been recently discovered that these fats can markedly increase LDL cholesterol and lower healthy HDL (high-density lipoprotein cholesterol). In response to research that conclusively shows a link to a diet with high trans fats and increased risk of cardiovascular disease, the U.S. Food and Drug Administration (FDA) announced in June 2015 that it will no longer allow trans fats in our food supply, and by 2017 it will be unlawful to have any in foods manufactured in the United States.

Healthier Fats

In general, healthier fats are mostly unsaturated fats. But as with all foods, too much of anything can have negative consequences. The "good" fats discussed below are found in a wide array of mostly plant and seafood sources.

Monounsaturated fats (MUFA) are found in a variety of foods and oils, but most commonly consumed as olive, canola, peanut, safflower, and sesame oils. Eating foods rich in monounsaturated fats has been shown to improve blood cholesterol levels, benefit insulin levels, and benefit blood sugar control.

Polyunsaturated fats (PUFAs) have similar benefits as monounsaturated fats, but because they vary significantly in how they are consumed, it is important to divide them into two groups (omega-3 and -6).

Saturated versus Unsaturated

Fatty acids can be grouped into two categories: those whose hydrocarbon chains are fully saturated with hydrogen atoms and those whose chains are not. Carbon atoms have only four "open" bonds available. In a fatty-acid hydrocarbon chain, at least two of those bonds connect each carbon atom to adjacent carbon atoms. When the remaining two bonds for all the carbon atoms in the chain are uniformly filled with hydrogen atoms, a fatty acid is "saturated." Fats made of saturated fatty acids are solid at room temperature.

If a carbon atom in the hydrocarbon chain is double-bonded to an adjacent carbon atom, however, leaving just one bond available for hydrogen, it is "unsaturated," because a hydrogen atom is now missing from the hydrocarbon chain. Fats made of unsaturated fatty acids are liquid at room temperature. An unsaturated fatty acid containing just one double bond is called monounsaturated. Fatty acids that have multiple double bonds are called polyunsaturated.

Understand Omega Fats

The two most commonly recognized dietary fats are omega-3 and omega-6. Despite their similar sounding names and structure, they can have widely different food source, functions and benefits through our body especially when it comes to disease prevention. In this section, we will compare and contrast their various sources and functions and explain why these two fats have become widely out of balance in our typical Western diet.

Omega-6 is found most abundantly in corn, soybean, safflower, sunflower, and canola oils, along with most nut-based oils. Both omega-6 and omega-3, the other type of PUFA, are also called essential fats (EFAs) since our body cannot produce them and they must be obtained from our diet. AA (arachidonic acid) is one of the most common omega-6s that is also essential for our health. Too often, however, our Western diet becomes overloaded with AA-containing foods that can have serious negative health consequences. We will discuss later why the balance of omega-3 to omega-6 is so critical to our health and why shifts in our greater consumption of omega-6 can contribute to inflammation and chronic diseases.

Omega-3 PUFAs are much less common in our diet and are found mostly in seafood. Unlike saturated fats, PUFAs remain liquid at both room temperature and in extreme cold. Omega-3 PUFAs are used by sea-dwelling animals to avoid freezing in the depths of the ocean. Omega-3s are also found in flax seed oils, certain green plants, and meats of animals that consume mostly a grass diet. The common omega-3s are EPA (eicosapentaenoic acid) and DHA (docosahexaenoic acid) and are mostly concentrated in seafood sources. ALA (α-Linolenic acid), a plant-based omega-3, is found in seeds like chia, flaxseed, and nuts. Both EPA and DHA are found in fish oil.

Both omega-3s EPA and DHA found concentrated in fish oil are now shown to benefit a wide range of health preservation and prevention roles. For example, DHA in particular is required for optimal neuronal (brain cell) function. A reduction in our dietary intake of omega-3 either as fish, seafood, or as a supplement has been linked to greater incidence of depressive illnesses. This same observation is true with cardiovascular disease as is shown in this table.

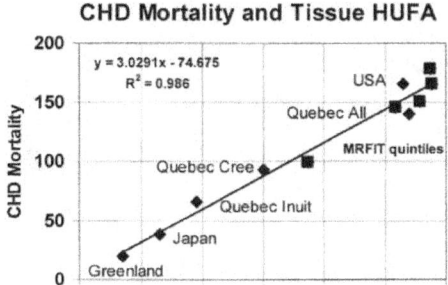

* HUFA: Highly unsaturated fatty acids. CHD (Coronary Heart Disease) Lipids, Vol. 38, no. 4 (2003)

Omega-3 vs. Omega-6

Research studies of our Paleolithic ancestors has suggested that human beings evolved consuming significantly less omega-6 fatty acids, and more omega-3 than is consumed today. As a result, an imbalanced ratio of omega-6 to omega-3 has occurred that is linked to the modern epidemic of CAD (cardiovascular disease.)

Western Diet Shifting Ratio of Omega-6 to Omega-3

Humans evolved from a diet of **1:1** (O-6 to O-3)
to approximately **10:1** to **20–25:1** seen today

As saturated fats and omega-6 increased so did CAD (Coronary Heart Disease)

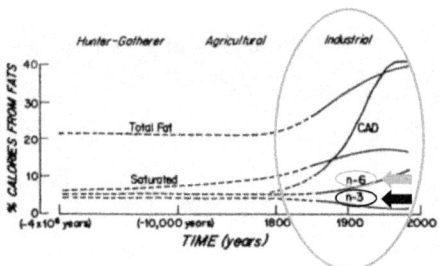

Exp.Biol.Med. 2008, 233,674-688

Vegetable-based oils, rich in omega-6 fats, were generally not available in abundance during much of human evolution but now dominate the modern Western diet. In fact, over the past one hundred years or so, we have raised the ratio of omega-3 to omega-6 EFAs in our diet to possibly as high as one to twenty-five.

There is also convincing research showing the positive effects of omega-3 fatty acids in hypertension, breast, prostate, and colon cancer, and skin and eye disorders, as well as many conditions affecting brain function and development. Adequate amounts of EPA and DHA in our diet can modify or reduce the inflammatory response associated with all of these conditions, as well as the aging process itself. By reducing inflammation-related damage to blood vessels, brain, skin, and other organs of the body, fish oil can have a wide range of beneficial effects.

American Heart Association Fish Oil Recommendations

TABLE 5. Summary of Recommendations for Omega-3 Fatty Acid Intake

Patient Population	Recommendation
Patients without documented CHD	Eat a variety of (preferably oily) fish at least twice a week. Include oils and foods rich in -linolenic acid (flaxseed, canola, and soybean oils; flaxseed and walnuts)
Patients with documented CHD	Consume 1 g of EPA/DHA per day, preferably from oily fish. EPA/DHA supplements could be considered in consultation with the physician.
Patients needing triglyceride lowering	Two to four grams of EPA/DHA per day provided as capsules under a physician's care

Circulation November 19, 2002 DOI: 10.1161/01

Healthcare Professionals Know the Benefits of Dietary Supplements over Drugs

- Healthcare Professionals Impact Study (HCP Impact Study 2009) found that **72% of physicians and 89% of nurses** used dietary supplements.
- **79% of physicians and 82% of nurses** recommend dietary supplements to their patients.
- There were 117,752 **deaths and** 711,232 **serious outcomes** in 2013 due to FDA approved drugs according to the FDA's Adverse Event Reporting System for prescription drugs.

Due to the extensive benefits of omega-3 EFAs, in 1996 the American Heart Association (AHA) first began recommending eating fish, especially fatty fish, at least twice a week. These recommendations were reaffirmed in 2002. The AHA also recommends eating tofu and other forms of soybeans, walnuts, flaxseeds, and canola oil in order to obtain essential amounts of omega-3. These plant sources contain alpha-linolenic acid (ALA), which is converted to omega-3 fatty acid in the body.

The father of medicine, Hippocrates, is credited with saying, *"utterly reject harm and mischief,"* when treating a patient. This moral obligation continues to today. Healthcare professionals are taught early on *"to do no harm"*. Yet, year after year, hundreds of thousands of people are injured and thousands kill unintentionally due to side effects and complications from pharmaceutical drugs provided by good intentioned healthcare professionals. The graph and table below show that this problem is now skyrocketing, with medical errors being the third most common cause of death in the U.S. Using natural supplements and treatments, like omega-3, for appropriate patients and their conditions, is not only preferred by the patients, if embrace by healthcare providers could begin to reverse this trend.

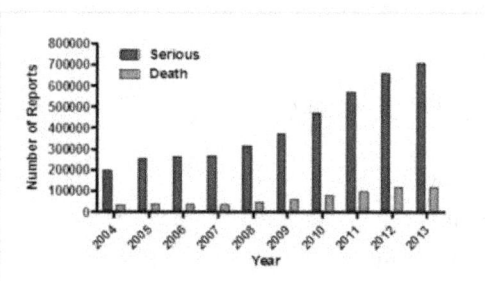

This graph and table represents the patient outcome(s) for reports since the year 2004 until the end of 2013. Serious outcomes include death, hospitalization, life threatening disability, congenital anomaly and/or other serious outcome due to prescription medications.

Year	Deaths	Serious Outcomes
2004	34,739	198,828
2005	40,031	256,208
2006	37,313	264,240
2007	36,689	272,345
2008	49,711	318,565
2009	63,842	373,512
2010	82,729	471,327
2011	98,590	573,402
2012	118,444	661,480
2013	117,752	711,232

CHAPTER

2

Inflammation is Good and Bad

To appreciate the benefits of omega-3s you must first understand inflammation. You could safely say that inflammation keeps us alive. Inflammation is our ultimate defense mechanism protecting us from a hostile world. Inflammation is produced by our immune system to ward off infections, allergies, and the ravages of disease. However, under certain conditions the inflammatory process can go unchecked, persist long after the original stimulus, and cause damage to the tissue it is trying to protect.

Local Inflammation

This is the kind of inflammation you've no doubt experienced firsthand, resulting from minor injury or infection. Take "pizza mouth" for example. Bite down on that hot mozzarella and the roof of your mouth quickly becomes red, hot, and very sore. Or fail to thoroughly cleanse and treat the finger you cut while gardening, and the next day the wound site is red, hot, swollen, and likely to be throbbing with pain.

Typically these symptoms respond positively to simple treatment such as applying ice, first aid ointment, and an over-the-counter pain reliever. Once the injuries have healed, typically there is no further damage or cause for concern.

Systemic Inflammation is Related to Heart Disease

The word systemic in medicine, means affecting the whole body. A systemic condition like vascular disease, diabetes and infections can cause almost every organ system in the body to suffer. The main destructive mechanism for many of these common chronic diseases is due to systemic inflammation and the havoc it causes to our cells. Only recently have scientist discovered that our unhealthy, high saturated fat and carbohydrate diet can also be the cause a systemic inflammation. This dietary link to inflammation may also be the cause of or worsen many other chronic conditions like Alzheimer's disease. Systemic inflammation caused by a poor diet and lifestyle is also known as silent inflammation (since we cannot feel it), and only often after many years and decades can the damage be seen. Systemic inflammation is related to heart disease. One of the most common diseases humans suffer from and the leading cause of death in the United States, heart disease and vascular disease are in part due to an inflammatory response. Atherosclerosis, commonly called hardening of the arteries, is caused in part by an inflammatory reaction after an invasion of lipids from the blood stream into the wall of the blood vessel itself. Once these lipids enter the inner lining (intima) of the artery, a "silent" inflammatory reaction occurs which results in more and more lipids entering the vessel. It is silent because we feel nothing despite the erosion of the artery. A plaque of calcium forms, the vessel starts to narrow, and if not detected can lead to stroke, heart attack, and high blood pressure.

Inflammation can also occur in overworked joints and causes the common arthritic conditions that most people have with aging. Although not as common, autoimmune diseases such as lupus and rheumatoid arthritis are caused by our immune inflammatory system attacking our own cells. It can be compared to friendly fire in a war, and in the same way results in tremendous suffering and pain.

Inflammation of the gastrointestinal tract, where seventy percent of our immune system is located, is another site where inflammation-related conditions such as Crohn's disease and irritable bowel syndrome (IBS) occur. Even emotional and physiological stress can result in an inflammatory response. Hormones such as cortisol, which remain high during times of chronic stress, can result in changes in the lining of the GI tract leading to increased risk of ulcers and gastritis. Over time, all those things that normally pass through or reside in the gut such as food molecules, bacteria, viruses, yeast, and toxins can leak through the intestinal wall and enter the bloodstream. This is referred to as "the leaky gut syndrome" and leads to an auto (self) immune reaction in which the body turns on itself.

For some people, consuming certain substances can have the same effect on the GI tract. Chances are you know people who react negatively to dietary gluten, lactose in milk, or genetically modified foods. Some people react to the presence of herbicides on fruits and vegetables. One of everybody's favorite "villains," sugar, can lead to the production of molecules that actually bind to hemoglobin and then circulates throughout the body. These molecules are called advanced glycosalated end products (AGEs). They are extremely inflammatory to tissues from arteries to skin, and are the cause of many of the common complications associated with diabetes.

As you might expect, some environmental pollutants, and even some products we use daily, can trigger the immune system. Fire retardants, cigarette smoke, some plastics, as well as the ingredients found in some cosmetics have been implicated as the source of an inflammatory response.

Eat more of these foods	Eat less of these foods
Whole grains (such as whole wheat bread, oatmeal, roti, and brown rice)	Refined or "enriched" grains (such as white bread, white rice, and white pasta)
Fish, lean meat, and poultry	Processed meats (including deli meats, bacon, sausages and hot dogs)
Calcium-rich foods (including low-fat milk, yogurt, and cheese)	High-fat foods (such as butter, cream, ice cream, and heavy cheese)
Unsaturated fats (from vegetable oil, nuts, and seeds)	Trans fats (found in processed food, cookies, cakes, and deep-fried food)
Vegetables and fruit	Sugary foods and drinks, especially processed desserts and both soda and diet sodas
Legumes (such as beans, peas, and lentils)	

Inflammation: Medicine vs. Omega-3 Fish Oil

Since we are constantly dealing with symptoms related to inflammation, savvy pharmaceutical companies have stepped up to fill this need. Chronic pain due to arthritic overused joints, headaches from vascular inflammation, or straightforward tissue injury motivates many of us to go to the medicine drawer for the anti-inflammatories we feel we can't live without. It is not surprising that anti-inflammatory medications are the largest group of medicines by volume sold throughout the world.

Blocking inflammation in the short term can not only help with pain relief, but can also help to stop further inflammation-related breakdown of cartilage and muscle in the case of overused joints. Unchecked inflammation can lead to tissue deformities and may even lead to genetic changes linked to cancers and disorders of the brain such as Alzheimer's disease. Understanding and treating the inflammatory response is the subject of thousands of medical papers, books, and symposia.

But what if the source of the irritation causing the inflammatory response persists? Can those same anti-inflammatory medications still do the job? The answer is "NO" for many people since long-term use of these medications can spell trouble. Despite being designed and tested for only short-term use, physicians prescribe and some people take these medications for decades! This is where history tells us again and again that there are unforeseen consequences to excessive use of painkillers despite the short-term relief they can provide.

Vioxx®: The Cautionary Tale of Anti-Inflammatory Medications

Despite a huge effort, billions of dollars spent, and extensive testing, serious complications can and do occur with prescription medications. One such example is the U.S. Food and Drug Administration (FDA) approval and use of the prescription drug Vioxx®. Known as a non-steroidal, anti-inflammatory drug (NSAID), it was heralded as a major breakthrough, and led to a completely new class of medications that were thought to have fewer stomach related problems and yet provide potent anti-inflammatory pain relief. A problem was discovered several years later however when researchers found a greater incidence of stroke and heart attack in those that were using this medication, often daily for months and even years. Further research showed that despite the potential for fewer stomach problems, the medication overtime increased the risk for development of blood clots that led to much more serious complications, including death. This medication, and several similar ones, was eventually pulled from the market.

This tragic event stimulated research into natural food supplements and plant sources, already shown to effectively reduce levels of pro-inflammatory factors, but with significantly fewer side effects. Sales of omega-3s, one of the most researched supplements in the world, known to provide significant inflammation reduction without side effects, began to skyrocket, and continues to grow every year.

Vioxx® is a registered trademark of Merck & Co.

NSAID Facts – American College of Gastroenterology

- NSAIDs are association with mucosal injury to the upper gastrointestinal tract, including the development of peptic ulcer disease and its complications, most notably upper gastrointestinal hemorrhage, and perforation.
- As many as 25% of chronic NSAID users will develop ulcer disease
- These gastrointestinal events result in more than 100,000 hospital admissions annually in the United States and between 7,000 and 10,000 deaths
- Risk factors for NSAID-related GI complications include a previous GI event, especially if complicated, age, concomitant use of anticoagulants, corticosteroids, other NSAIDs including low-dose aspirin, high-dose NSAID therapy, and chronic debilitating disorders, especially cardiovascular disease.
- Low-dose aspirin is associated with a definite risk for GI complications.
- H. pylori infection increases the risk of NSAID-related GI complications.

Reference: Lanza, FL., Chan, FKL, Prevention of NSAID-Related Ulcer Complications, Am J Gastroenterol 2009; 104:728–738; 2009.115;

The FDA in July 2015 also strengthened its existing warning in prescription drug labels and over-the-counter (OTC) drugs. To read:

> "Facts labels to indicate that nonsteroidal anti-inflammatory drugs (NSAIDs) can increase the chance of a heart attack or stroke, either of which can lead to death. Those serious side effects can occur as early as the first few weeks of using an NSAID, and the risk might rise the longer people take NSAIDs. (Although aspirin is also an NSAID, this revised warning doesn't apply to aspirin.)"

Reference: http://www.fda.gov/downloads/Drugs/DrugSafety/ucm089162.pdf

In addition, the FDA warns that NSAID medicines may increase the chance of a heart attack or stroke that can lead to death. Also, NSAID medicines should never be used right before or after bypass heart surgery. Perhaps most importantly you should know that NSAID medicines should only be used exactly as prescribed, at the lowest dose and for the shortest time needed.

Even Physicians CAN Change!

Because of the failure of Vioxx®, physicians began to rethink their use of NSAIDs such as ibuprofen and naproxen. Many, including me, became convinced, through both the powerful research evidence and personal use, of just how effective some natural supplements such as fish oil can be. The use of natural supplements for inflammatory-related diseases, *and* even disease prevention, has opened a whole new perspective on how to keep people healthy.

To understand the benefits of omega-3s it is important to know how it is, and is not, like NSAID medications. Structurally, omega-3 fatty acids found in seafood and concentrated into fish oil have two major active ingredients: EPA (eicosapentaenoic acid) and DHA (docosahexaenoic acid). Since the 1970s, when the first scientists discovered the beneficial mechanisms of EPA and DHA, there have been more than seven thousand scientific reports, including thousands of human clinical trials, showing the many human health benefits. Omega-3 use in cardiovascular disease, stroke, cancer, hyperlipidemia, Alzheimer's disease, attention deficit hyperactivity disorder (ADHD), rheumatoid arthritis, and many others have all shown remarkable benefits.

Regular use of fish oil supplements lowers the body's levels of the inflammatory factors, thereby blunting the inflammatory response like NSAIDs do. However, unlike certain prescription NSAIDs, known as selective COX-2 inhibitors, omega-3s in fish oil have not been found to increase clotting risk, nor result in stomach ulcers and GI bleeding. In fact, Celebrex™, a prescription NSAID selective COX-2 inhibitor, actually has a "Black Box" warning (required by the FDA when there is reasonable evidence of an association of a serious hazard with the drug) on its label due to increased risk of heart attacks, stroke, and GI bleeding (see below.) Overall the benefits of fish oil and its almost non-existent side effects can provide most people benefits well beyond NSAIDs.

WARNING: RISK OF SERIOUS CARDIOVASCULAR AND GASTROINTESTINAL EVENTS

See full prescribing information for complete boxed warning.

Nonsteroidal anti-inflammatory drugs (NSAIDs) cause an increased risk of serious cardiovascular thrombotic events, including myocardial infarction and stroke, which can be fatal. This risk may occur early in the treatment and may increase with duration of use. (5.1)

CELEBREX is contraindicated in the setting of coronary artery bypass graft (CABG) surgery. (4, 5.1)

NSAIDs cause an increased risk of serious gastrointestinal (GI) adverse events including bleeding, ulceration, and perforation of the stomach or intestines, which can be fatal. These events can occur at any time during use and without warning symptoms. Elderly patients and patients with a prior history of peptic ulcer disease and/or GI bleeding are at greater risk for serious GI events. (5.2)

Reference: Celecoxib capsule labeling by G.D. Searle LLC Division of Pfizer

CHAPTER

3

A Closer Look at EPA and DHA

The most important and bioactive omega-3 fatty acids are EPA and DHA. Typically most fish oil supplements contain a higher concentration of EPA than DHA. This may be due to manufacturer costs. Seafood has both EPA and DHA fairly equally distributed. Research investigating the roles and health benefits of both EPA and DHA separately, and in conjugation, has revealed a variety of different benefits throughout our body.

DHA omega-3 fatty acids are found to be extremely important to healthy brain function. EPA offers other benefits such as greater reduction in inflammation, which can play a roll throughout the body. Both EPA and DHA also have similar effects despite playing separate roles in contributing to better health. This has led scientists and manufacturers to look closer at the ratio of EPA to DHA in an effort to find the ideal fish oil supplement.

EPA

Recent research suggests that EPA and DHA actually serve different functions in promoting health, but both are needed for a wide variety of benefits. EPA seems to play a larger role in reducing cellular inflammation than does DHA. We'll be talking about arachidonic acid (AA) a little further on, but for the time being accept that AA is a natural molecule, a type of omega-6 polyunsaturated fatty acids (PUFA), found in most vegetable oils in our diet. Unfortunately too much AA, as a result of our diet choices, can be used as a building block for the production of inflammatory molecules.

One of EPAs major functions is to compete with the overproduction of AA inflammatory molecules in order to counter excessive inflammatory reactions throughout the body. For many people the greater amount of EPA in a person's diet, the less AA related inflammation could occur. The goal is to seek a healthy balance between EPA and AA. Unfortunately our modern diet is overwhelmed with too many AA sources, such as the vegetable and nut oils found in most processed foods we eat.

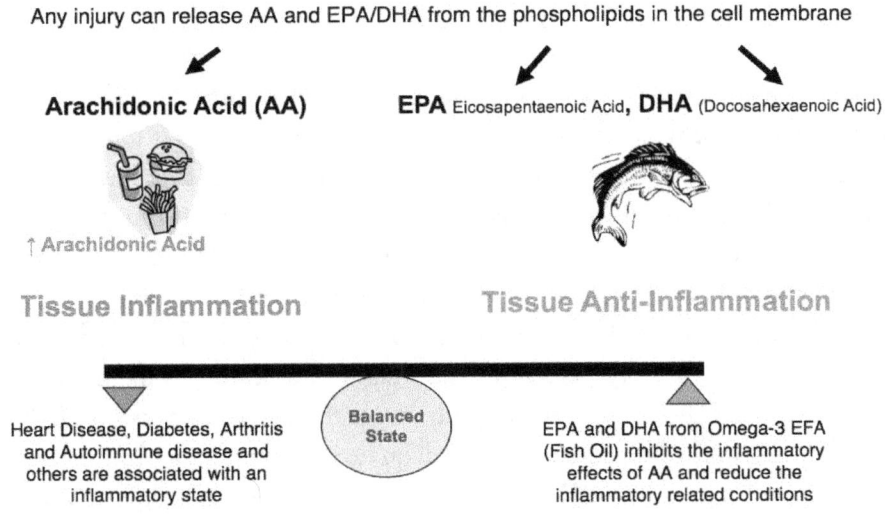

Cell Membrane Injury

Any injury can release AA and EPA/DHA from the phospholipids in the cell membrane

Arachidonic Acid (AA)   **EPA** Eicosapentaenoic Acid, **DHA** (Docosahexaenoic Acid)

↑ Arachidonic Acid

Tissue Inflammation   Tissue Anti-Inflammation

Heart Disease, Diabetes, Arthritis and Autoimmune disease and others are associated with an inflammatory state

Balanced State

EPA and DHA from Omega-3 EFA (Fish Oil) inhibits the inflammatory effects of AA and reduce the inflammatory related conditions

Figure: Balance Between Omega-3 and Omega-6

Fish Oil Has
Both EPA and DHA

There are lower levels of EPA than DHA in the brain, yet EPA is still important for healthy brain function. It is key for reducing neuro-inflammation by thwarting excessive pro- inflammatory AA. The problem is that EPA is not stored in the brain and can rapidly become depleted, unlike DHA which is incorporated into every brain cell membrane. Therefore, many physicians believe that for brain health, more EPA than DHA is needed. Manufacturers have responded by producing higher levels of EPA in their fish oil capsules.

Greater levels of EPA have also been shown to help patients with brain trauma due to a greater anti-inflammatory effect. In a clinical trial, patients taking omega-3 supplements with higher ratios of EPA to DHA experienced fewer depressive symptoms. Two studies using pure EPA showed remission in patients with Treatment Resistant Depression. Two grams daily of DHA by itself had no effect on mood in a group of patients with major depression.

DHA

DHA is another type of omega-3 essential fatty acid, but is not known to play a major role in inflammation reduction. DHA has been found to be especially important for our brain health and other tissues that have higher concentrations of nerve cells. It is not a coincidence that fish and seafood are called "brain food." DHA, found in abundance in seafood, is the most prevalent omega-3 fatty acid in the brain and retina for this reason. Up to forty percent of PUFAs in the brain and sixty percent of PUFA in the retina are made of DHA. Fifty percent of the weight of a brain cell membrane is composed of DHA. DHA is structurally larger than EPA. Because of DHA's unique structure, when consumed in adequate amounts, it can improve the function of brain cell membranes. The specific benefit is called increased fluidity. This simply means the improved ability of brain cells to "talk" to each other. Increased cellular fluidity is important for both brain cells and the retina of the eye because it is the action of transmission fluidity that allows for greater signal transfer. Thinking faster and better is clearly an advantage we all could benefit from. This is also why DHA has been studied for reducing cognitive decline and Alzheimer's disease symptoms.

DHA, generally obtained from prenatal fish oil supplements, are crucial for fetal brain development. Since children are not as easily able to metabolize DHA and EPA from ALA, the plant source of omega-3, fish oil supplements are more often recommended. Infants of mothers supplemented with DHA during pregnancy and while nursing have been reported to have higher mental processing scores and psychomotor development, better eye-hand coordination and visual acuity at four years of age compared to infants not supplemented. Intake of DHAs during preschool years may be beneficial in the prevention of ADHD, as well as enhancing learning capability and academic performance.

Adequate amounts of DHA in cell membranes have been shown to better help resist cancer and make it more difficult for inflammatory molecules to initiate the inflammatory response.

Both EPA and DHA are effective at lowering triglyceride levels. Elevations in triglycerides are frequently linked to carbohydrate or sugar consumption, as well as linked to increased risk of cardiovascular disease. Although controversial, large doses of DHA found in certain fish oil based dietary supplements can result in increases of LDL on a cholesterol blood test. The significance of this has been debated, but higher levels of LDL are a known risk factor for atherosclerosis, commonly known as hardening of the arteries. Based on this possible problem, some manufacturers have begun to lower the DHA content and increase the amount of EPA in their omega-3 fish oil supplements. Later on in this book we will further review the importance of EPA to DHA and the latest research on this topic.

Summary:

Both EPA and DHA are important for maintaining a healthy body and brain. EPA appears to be more effective at reducing inflammation. A large number of physical and mental health conditions are linked to excessive inflammatory response. These include cardiovascular disease, arthritis, cancer, stoke, and autoimmune diseases of the gut. Mental health conditions including depression, ADHD, mood disorders, and even Alzheimer's disease have inflammatory affects for which EPA has shown a benefit.

DHA is effective at promoting brain cell communications and may slow cognitive decline, macular degeneration, and cardiac electrical dysfunction. DHA plays a more important role in brain and retina development, especially during pregnancy and infancy.

So What is the Optimal Ratio of EPA to DHA?

Not until recently has much thought been given to how much EPA or DHA should be consumed either in our diet or as dietary supplements. Since both have major health benefits and both are found mixed together in food sources like seafood, the absolute amounts of each didn't seem to matter. This oversight was reversed as the popularity of dietary supplements, in particular fish oil supplements, has exploded throughout the world. There are now hundreds of scientific studies published every year promoting the many health benefits of omega-3, and the majority are using dietary supplements with a variety of EPA to DHA ratios.

The scientific evidence about the benefits of fish oil is complex when it comes to the best ratios of EPA to DHA. However, some general trends can be seen based on organ or condition-specific benefits. Using the various characteristic differences and similarities of EPA and DHA for human health, scientists who have studied and published on omega-3s generally use dramatically different ratios of EPA to DHA. Manufacturers and developers of fish oil supplements use all these facts, including cost and purity, in order to make fish oil supplements with unique health benefits.

For example, macular degeneration and other conditions of the eye have been shown to benefit when supplementing with omega-3 fish oil. The eye is known to have a higher concentration of DHA. This fact, along with scientific studies that show more DHA containing supplementation has a greater eye health benefit, has prompted manufacturers of fish oil supplements to promote eye health to make their products with higher ratios of DHA to EPA. This area of the manufacturing process is referred to as formulation. Most people who are told to take fish oil by a physician, or have learned on their own about the benefits of fish oil supplements, generally buy fish oil supplements without knowing these important facts on EPA to DHA ratios.

This is a simple list made for three common health areas where omega-3s are known to benefit. Also listed are suggested EPA to DHA ratios that may provide the most benefit:

Cardiac Health (More EPA/Less DHA)

- More EPA (anti-inflammatory, lower triglycerides) vs. DHA (needed for heart rhythm function, but not too high, otherwise it will increase LDL (bad cholesterol)

Brain Health (similar amounts of EPA to DHA)

- EPA (more needed to cross blood brain barrier) vs. DHA (more needed for structure of brain cell membranes)

Joint Health (More EPA/Less DHA)

- More EPA (anti-inflammatory) vs. DHA

CHAPTER

4

Understanding Fish Oils Supplements

So with all this knowledge and research, what is the ideal ratio of EPA to DHA for a fish oil supplement? This is the question that perhaps hundreds of fish oil formulators and millions and millions of Americans ask every year. For the reasons listed above, the ideal dose and ratio of EPA to DHA in a fish oil supplement comes down to why you want to take fish oil in the first place. Because most Americans are deficient in both EPA and DHA, and omega-3 in general, almost any fish oil supplement can help to increase these levels and provide some benefits. There are some exceptions to this statement that will be reviewed in the fish oil quality section.

Fact: Americans are Starved of Nutrients!

1. According to National Institute of Health, the three most common foods consumed by adult Americans based on calories are: desserts, breads, and chicken products, with potatoes (a vegetable) being 16th.

2. A survey in 2011 by *Consumer Reports* showed one in three adults is obese, and only about 1/3 of Americans surveyed eat the recommended five or more servings of fresh fruit or vegetables daily. Why?
    a. They thought they consumed enough already.

---

    b.   Fresh fruit and vegetables are too hard to store or spoil too quickly.

    c.   Household didn't like vegetables, take too long to prepare or are too difficult to prepare, and fresh vegetables are too expensive.

3.   *The International Food Information Council Foundation's* 2012 Food & Health Survey found these have actually gotten worse:

    a.   Only *sixty percent* of Americans say their health is either excellent or very good.

    b.   Only *twenty-five percent* report eating a healthy diet.

    c.   Only *fifteen percent* accurately estimate calories to maintain weight.

    d.   Only *six percent* thought about their physical activity.

So, what if you are concerned about heart disease? What is the best ratio of EPA to DHA for you? Unfortunately, heart disease is still the number one cause of death in the United States. The risk factors are all around us, and in us. The toxins in our environment, smoking, excessive alcohol, lack of regular exercise, genetic factors, and our generally poor diet are causative.

One of the most identifiable factors linked to heart disease are abnormal cholesterol levels. Cholesterol, also called lipids, comes in a variety of types. These lipids are important fats used to make our cell membranes, especially in our brain, and are important as the backbone molecule for various hormones in our body.

Triglycerides are less often talked about, but are another type of lipid that is just as important for heart health. At normal levels, triglycerides are used to transport extra fats from our food and stored as body fat (also known as adipose tissue) to be used later as energy. Triglycerides can also be used in reverse to transport stored fats to the liver to make new energy. Of course, too many of us know what happens if we don't use that energy. It remains stored and we gain weight. Excess triglycerides are converted to fat and are deposited in our belly, buttocks, and legs. Triglycerides are also a serious heart health risk and unfortunately, unlike dietary changes, not only do most prescription drugs do little to lower triglyceride levels, they also have numerous side effects.

Your doctor routinely checks cholesterol and triglycerides levels. There are now specialized tests for cholesterol that can determine all the various types including those found to have more detrimental heart effects. One of these tests is called Vertical Auto Profile Test (VAP). The table below describes each type of cholesterol and the desired levels. There remains a great deal of controversy in the cardiology world as to the appropriate cholesterol value in order to reduce heart disease risk.

Table 2: Cholesterol Subtractions, Risk Description and Normal Values

Cholesterol			
Type	Description	Major Components	Desirable Score
Total Cholesterol	Total cholesterol circulation in body.	all cholesterol	<200 mg/dL
LDL Cholesterol	Considered to be the "bad" cholesterol because it is a primary cause of heart disease.	Lp(a), IDL, LDL-R	<130 mg/dL
HDL Cholesterol	Considered the "good" cholesterol because low levels of this can lead to heart disease.	HDL2, HDL3	>=40 mg/dL
VLDL Cholesterol	Carrier for triglycerides. If high it is a risk for heart disease.	VLDL3	<30 mg/dL
Triglycerides	Molecules that provide energy to the entire body. If levels are high, triglycerides are a risk for heart disease.	Several	<150 mg/dL
Non-HDL Cholesterol	Contains all bad cholesterol components and subclasses LDL and VLDL.	several-see description	<160 mg/dL
Cholesterol Sub-fractions			
Type	Major Components	Description	Desirable Score
LDL Cholesterol	Lp(a)	Very dangerous cholesterol that is harder than most to treat effectively with drugs.	<10 mg/dL
	IDL	Dangerous if elevated.	<20 mg/dL
	Real LDL Cholesterol	Component of LDL cholesterol, measures the real cholesterol circulation in the body.	<100 mg/dL
	LDL Cholesterol Pattern Size	LDL cholesterol ranges from small and dense (Pattern B) to large and buoyant (Pattern A). The smaller the LDL size, the greater the risk for heart disease.	A
HDL Cholesterol	HDL2	The most protective form of HDL, large and buoyant	>10 mg/dL

Reference

Sports
Health. 2009

One of the most common ways cardiologists lower cholesterol levels is by using a type of drug called statins. The use of statins to lower cholesterol is at an all-time high in the United States and around the world. Statin prescriptions in the United States during the years 2007 to 2012 grew nearly twenty percent per year, and the trend is accelerating. Total global sales in 2012 were $35 billion including $10 billion in the United States. In 2012 the Centers for Disease Control (CDC) reported more than seventy-three million adults in the United States have high LDL (the bad cholesterol). These patients have twice the risk of heart disease as compared to those with normal or low LDL. Barring most genetic conditions that markedly increase cholesterol levels, most people respond to medications to help lower the cholesterol levels. Nevertheless, pills alone are not a panacea for lowering cholesterol and heart disease risk. A healthy diet, not smoking, and regular exercise are also critical factors for heart health.

So to answer the original questions about fish oil and heart health, omega-3s can have a profound benefit for our cardiovascular system by reducing inflammation in our blood vessels and lowering lipids in our blood. As mentioned in the beginning of the book, the Inuit Eskimos, despite a high-fat diet, generally were found to have a lower incidence of heart and blood vessel disease. Much of this benefit was actually believed to be due how omega-3s can control triglyceride levels. This finding was certainly not expected, because most people once believed that any fat in our diet would result in elevation of blood lipids. Additional studies have confirmed similar findings in other populations.

The first prescription fish oil in 2009, Lovaza®, received FDA approval based on its ability to lower triglyceride levels. Lovaza as a prescription fish oil has done very well with over $1 billion in annual sales. The EPA to DHA ratio of Lovaza is 1.2 to 1. (**See Comparison Table**) Looking at this ratio you can see that DHA is only somewhat lower than the EPA amount. As discussed, both EPA and DHA have heart health benefits. However, early studies using Omacor™, the precursor to Lovaza® but with the same formulation, revealed significantly increased LDL (the bad cholesterol). In fact a thirty-one percent increase in LDL was reported.

Reference

AMJ Cardiol 2006; 98 (supp); 71-76

Lovaza® is a registered trademark of GlaxoSmithKline

What is the Best EPA to DHA Ratio in a Fish Oil Supplement?

Company	Product	EPA+DHA (mg)	EPA(mg)	DHA(mg)	Ratio(EPA:DHA)
GSK	Lovaza	3360	1860	1500	1.2:1
Fisol	Fish Oil	250	150	100	1.5:1
GNC	Triple Strength Fish Oil 1500	900	540	360	1.5:1
GNC	Triple Strength Fish Oil 1400+ Brain Support	900	647	253	2.6:1
GNC	Triple Strength Fish Oil 1400+ Joint Support	900	647	253	2.6:1
MusclePharm	Core Series Fish Oil	700	400	300	1.3:1
Natrol	Omega-3 Fish Oil 1000mg	300	180	120	1.5:1
Nature Made	Fish Oil Adult Gummies	57	9.5	47.5	1.0:5
Omax3	Omax3 Ultra-Pure	1400	1125	275	4.1:1
Renew Life Formulas	Norwegian Gold Ultimate Super Critical Omega	900	780	120	6.5:1

So what is the ideal amount of DHA in a fish oil supplement? As seen from the table, some fish oil supplements have very low levels of DHA. Since DHA is critical for brain, eye, and heart cells associated with a normal heart rhythm, can an all EPA product, or one with very low DHA, provide nutritional benefit for these organs? On the other hand, if a product has too much DHA compared to EPA, will this result in an excessive triglyceride increases due to a DHA overload? So what is the sweet spot when formulating EPA and DHA into an effective fish oil supplement? What are the most common health areas that people want to use fish oil supplements to help improve? The table below is a list of some of our most common health conditions and those where omega-3 studies have shown significant benefits. In the following chapters we will dive deeper into each of these areas to give you a fuller picture of our critical need to consume enough of this essential fatty acid.

Dig Deeper

What is the Best EPA:DHA Ratio?

One of the best ways to determine the most desirable EPA to DHA ratio for a particular fish oil supplement is for manufacturers and formulators to conduct studies, using both animals and humans, to test their proprietary formulations. Several manufacturers have found ratios of EPA to DHA that appear to benefit a wide variety of health concerns. Others seek to target a ratio to help one specific condition or organ to benefit. This was the approach that manufacturers of the prescription fish oil, Lovaza® (EPA to DHA is 1.2 to 1) took to obtain a very narrow FDA approval that targeted elevated triglycerides.

Recently a non-prescription fish oil supplement manufacturer published research using a patented ratio of four to one EPA to DHA that found significant and wide ranging health benefits. Some of these results are highlighted below:

- Showed significant anti-depressant properties in animals.
- Significantly increased a brain cell-building hormone called BDNF (Brain Derived Neurotrophic Factor) in animals.
- Significantly decreased inflammatory molecules in healthy people at a rate five times greater than fish oil supplements formulated using EPA to DHA ratios with greater amounts of EPA and others with greater amounts of DHA.
- In healthy people without abnormal lipids reduced triglyceride levels an average of thirty-one percent, cholesterol by thirteen percent, and LDL "bad cholesterol" by eleven percent.

> **Be careful!** Taking fish oil supplements can offer numerous health benefits, but fish oil—like all food or dietary supplements—can interact with certain medications and disease conditions. Talk to your doctor if you are just starting on fish oil supplements or are on any blood thinners (anticoagulants) or prescription medications. The majority of safety studies report fish oil is safe for most people when taken in doses of 3 grams (gm) or 3,000 milligrams (mg) or less daily. People who routinely consume a daily dose of more than 3,000 mg of fish oil, however, may have an increased risk for bleeding, a weakened immune system (especially the elderly), and nausea, heartburn, and nosebleeds. There are, however, various studies that have used up to 5 gm or even greater of mixed EPA/DHA fish oil reporting no side effects. Again, discuss with your healthcare provider possible side effects and interactions with your existing medical conditions and medications before using.

Quality of Fish Oil

When choosing a fish oil supplement, it is critical to find a high quality product that has been purified of potential toxins. Since fish oil is extracted from saltwater fish and seafood, toxins such as mercury, lead, and polychlorinated biphenyls (PCBs) due to pollution, must be extracted from fish oil supplements prior to being sold. This is a complex and expensive process and varies between fish oil processors.

Dig Deeper

Fatty fish like mackerel, lake trout, anchovies, herring, sardines, albacore tuna, and salmon can provide concentrated amounts of EPA and DHA. One of the major drawbacks of consuming fish as your major source of omega-3 is that it can have significant concentrations of pollutants stored in its flesh. Toxins concentrated up the food chain can become concentrated in fish. This is particularly true with larger fish like tuna, shark, and swordfish that eat other fish and live longer. Toxins like lead, mercury, PCBs (polychlorinated biphenyls), and dioxin, all have serious health consequences if consumed frequently.

Other often-unrecognized threats to the quality fish oil supplements are air, heat, and sunlight. Fish oil, is polyunsaturated oil that will begin to turn rancid if exposed to excessive heat or prolonged exposure to sunlight or air. This is the same process that can make fish smell if left out on a kitchen counter for too long. We all know this smell and if your fish oil supplement smells like rotting fish or seafood "gone bad," you should avoid taking it. In addition, burping after taking some fish oil supplements and the foul taste of others can also be related to rancidity and is a sign to avoid taking that particular supplement.

Quality fish oil supplement manufacturers are working hard to limit exposure to these environmental threats. When looking for the best fish oil supplement, inquire about how the fish oil is processed. Are they using an oxygen-free environment during manufacture to limit rancidity in their product? Do they avoid high heat and use bottles to sell their product that are resistant to sunlight? Do they use the purest fish oil with potentially toxic pollutants removed? And are they using the most concentrated amounts of EPA and DHA to avoid extra fats and contaminants in the oil? Certain fish oil suppliers have taken these concerns to a whole new level by enclosing each two-capsule dose of fish oil in a foil blister pack that further limits air exposure that would otherwise occur when opening and closing a bottle. In addition, some manufacturers use up to ninety-five percent pure EPA and DHA formulations that not only can make the capsule size smaller, but eliminate saturated and contaminated fats often found in less pure products.

What is the Best Fish Oil Dose?

To determine the best dose of fish oil supplement for you, first determine your goals. Do you mostly avoid fish and seafood and need to supplement your diet? Are your concerned about your cardiovascular health and uncontrolled inflammation? Have you been told you have elevated triglycerides and want a more natural way to help lower it? Do you want fish oil for brain health or for your mood? The good news is that even if you only believe you need fish oil supplements for a specific health concern, omega-3s can provide additional benefits for the whole body.

Earlier we reviewed the AHA recommendations for coronary heart disease (CHD). We recommend one gram of mixed EPA and DHA daily for history of CHD and two to four grams for elevated triglycerides. Again you must make sure to read labels on the fish oil supplements. If the total amount of EPA and DHA doesn't add up to the 1,000 mg recommended for CHD, then you must take as many capsules as needed to reach this amount. This is why concentrated fish oil supplements are so popular since then can necessitate fewer capsules to be taken each day.

In general, we recommend at least 1,000 mg of mixed EPA and DHA per day as a minimal beneficial dose for fish oil supplements. Many studies report benefits using dosages from one to three grams of EPA/DHA per day or even higher, based on certain conditions. In the next several chapters we will dive deeper into brain, emotional, joint, skin, eye, and intestinal health and provide you some of the best omega-3 human studies for these and other health benefits.

In summary, there is now strong scientific evidence that modern dietary shifts resulting in too little omega-3 in our diets have had profound health effects. The epidemic of inflammatory related diseases, such as heart disease, arthritis, cancer, and even Alzheimer's disease, are clearly linked to Western dietary changes, lack of exercise, and increased environmental toxic exposures. Supplementing with omega-3 fatty acids found in fish oil supplements can result in a substantial benefit for many regardless of age, sex, or current health condition. As healthcare practitioners, we feel an obligation to tell the public about natural food based supplements that have stood up to rigorous clinical testing and can offer an alternative to often expensive prescription medications and their potential side effects.

Now that we've discussed the general reasons why dietary and high quality omega-3 supplements are important to your overall wellness, let's take a look at how this essential fatty acid can affect specific aspects of health.

CHAPTER

5

Let's Take it From the Top: Brain Health

There are approximately one hundred billion nerve cells or neurons that make up the two-and-one-half to three-pound ink-gray gelatinous, fascinating, wonderful, mysterious, crucial, and miraculous organ we call the brain. The brain is the inspiration for the phrase "nerve center" because it truly does control every function of the human body. Nothing works without it; from all the things we do consciously, to all the activities the body performs automatically without our awareness. The brain allows us to think, feel, hear, smell taste, move, and see.

Using a series of complex electrical signals and chemical processes, the brain orchestrates all of the body's functions. However, in order to do this, cells within the brain, in a matter of milliseconds, must communicate with other brain cells to correctly direct where these signals go. The key to this communication system is the ability of brain cells to "talk" to each other. The "talking" occurs at synapses, the points where two brain cells touch each other. Each neuron has up to five thousand connections with other cells.

Simply explained, a synapse is an extremely small space between the connection points of nerve cells where the information is transferred and sent further along. Whether it is the nerve cell in the brain or elsewhere in the body, this is how information is transferred. And despite the many twists and turns and miles of nerves an impulse travels, it only takes a fraction of a second to reach its destination.

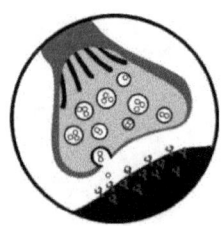

Nerve Cell Synapse

The key to omega-3s effect on the brain is on the synapses — the tiny gaps across which nerve impulses cross from one neuron to another. In order for the impulse molecules (called neurotransmitters) to reach the receiving neuron, they need to pass through a membrane that surrounds the neuron. Cell membranes are made almost completely of fats, including the omega-3 DHA. The DHA molecules help keep the membrane more elastic, enhancing the flow of electrical impulses and even making new nerve connections more easily.

The speed and efficiency of this process can be influenced by hundreds of different factors and ultimately determines the function of our brain. Some factors such as aging and underlying genetics cannot be changed, but factors like our diet, level of exercise, and environmental exposures can be changed in order to influence, preserve, and even improve brain function.

The omega-3 DHA is not only essential as a building block for the nerve cell, and it is one of the primary factors for the efficiency for synapses to communicate. Numerous studies have confirmed that the omega-3 fatty acids found in fish are essential for optimal brain function. People who don't get enough omega-3s in their diet have been found to have an increased risk of dementia, depression, attention-deficit disorder, dyslexia, and schizophrenia.

Omega-3s also help improve blood flow in the brain. Research is finding that adequate intake of omega-3 fatty acids has numerous brain-boosting benefits, such as:

- Improved learning and memory. In studies, children who received omega-3 fatty acid supplements did better in school, scored higher on tests of learning and memory, and had fewer behavioral problems than their peers who didn't get the supplement.

- They protect against depression and other mood disorders, schizophrenia, and improve mood in people who already have depression.
- They fight age-related cognitive decline due to dementia.

The benefits of omega-3s begin before birth as the DHA component is passed to the fetus across the placenta. After birth, omega-3s in breast milk help build new brain cells and also supports the retina of the developing eye. DHA is often added to infant formulas since, just like the rest of the population, pregnant women are often deficient in omega-3. One of the reasons for this is the FDA and the U.S. Environmental Protection Agency (EPA) both have had long standing recommendations for pregnant women and young children to restrict their fish consumption due to the associated toxins such as lead and mercury found in most fish and seafood. (See Side Bar)

Another study found that children of mothers who took fish oil supplements during pregnancy had higher IQs than those whose mothers took a placebo. Babies need DHA — especially during the first two years of life —for their brains to develop properly. One study found that babies who were born to mothers with higher blood levels of DHA scored higher on tests of attention and learning than those whose mothers had lower DHA levels.

Is Fish Safe?

Fish is an excellent source of protein and other important nutrients. If you're a woman who is pregnant or nursing, fish contains critical brain food, omega-3s that are essential for your baby's brain and eye development. Unfortunately the majority of fish and seafood that comes from both fresh and sea water has hidden and often untested dangers. Toxins, such as mercury and lead found in the air, and mostly from pollution due to power plants, are carried by the rain into the world's water supply and then are slowly concentrated in the flesh of fish and other seafood. In small amounts fish is generally harmless to humans, but excessive amounts can result in negative effects, especially for the brain.

In 2014, the FDA and EPA proposed new guidelines recommending that women of childbearing age and young children eat more fish and seafood. But the catch is to only eat more fish and seafood known to have lower levels of dangerous toxins.

There are lists available for these low-toxin fish, but the key question is, can Americans, especially pregnant women and children, identify these special fish and seafood types? Safety organizations such the Consumers Union, publisher of the magazine *Consumer Reports*, made the following comment:

"Though the (governmental) agencies say consumers should seek out fish that are low in mercury, almost all seafood contains the toxin (mercury) in varying amounts, and getting too much of it can damage the brain and nervous system. That is especially true for fetuses, but children and adults who eat too much high-mercury seafood also can suffer harmful effects such as problems with fine motor coordination, speech, sleep, and walking, and prickly sensations.

Consumer Reports' food-safety experts analyzed the FDA's own data that measures mercury levels in various types of seafood. From that we identified almost twenty seafood choices that can be eaten several times per week, even by pregnant women and young children, without worrying about mercury exposure.

However, Consumer Reports disagrees with the recommendations from the FDA and EPA on how much tuna women and children may eat. 'We don't think pregnant women should eat any.' We also believe the agencies do not do enough to guide consumers to the best low-mercury seafood choices."

Due to the popularity of canned tuna and shrimp, both known to have higher mercury content (canned tuna accounts for twenty-eight percent of Americans' exposure as per *Consumer Reports*), we also remain skeptical that this high-risk group can avoid excessive exposures. Fish oil supplements are and should be a critical part of prenatal care because they are essential for the developing brain of children.

Keys to a healthy brain

The remainder of this chapter will look at eight steps you can take in the area of nutrition that can directly benefit the gray matter in your head—your brain! These steps are:

(1) Lose excessive weight

(2) Address malnourishment

(3) Stay hydrated

(4) Get enough protein

(5) Choose healthy carbohydrates

(6) Avoid excess sugar

(7) Know the difference between "good" and "bad" fats

(8) Increase your intake of protective foods

Aging and Cognitive Impairment

Cognition is an all-encompassing term that includes mental processes such as thinking, reasoning, interpreting information, understanding, and memory. As individuals age, we are especially concerned about the effects of memory loss and a decline in the ability to reason. Diseases such as Alzheimer's and dementia have taken a tremendous toll on our elderly and continue to be a huge medical cost to society. We are a long way from fully understanding how and why aging changes our mental acuity, but progress is being made continually.

As the aging process unfolds in all of us, the brain — just like the rest of the body — goes through changes. Narrowing of arteries and reduced growth of capillaries can reduce the blood flow to the brain. Parts of the brain shrink, including those responsible for complex activities such as learning and memory. Communication ability between neurons and neurotransmitters can be reduced. Damage by dietary free radicals, due to excessive blood sugar, saturated fats, and chronic inflammation increases over time. As a result of aging, otherwise healthy individuals can begin to experience memory loss, difficulty in completing complex tasks, and the inability to master new skills or learn something new. Other factors that can contribute to mental decline must include our genetic blueprint, lifestyle, general health, and disease burden.

Although common with aging, brain decline does not have to occur. Some people have the ability to compensate, known as a cognitive reserve, but preserving brain function as we age is a process that should start long before symptoms of brain decline appear. Ideally, from birth to death, brain preservation should be the guiding principle for decisions on your level of activity, diet, stress control, and toxin exposure.

Alzheimer's Disease Facts

An estimated 5.2 million Americans were reported to have Alzheimer's disease in 2014, including approximately two hundred thousand individuals younger than age sixty-five who have younger-onset Alzheimer's. Almost two-thirds of American seniors living with Alzheimer's disease are women. Of the five million people age sixty-five and older with Alzheimer's disease in the United States, 3.2 million are women and 1.8 million are men.

The number of Americans with Alzheimer's disease and other dementias will escalate rapidly in coming years as the baby boom generation ages. By 2050, the number of people age sixty-five and older with Alzheimer's disease may nearly triple, from five million to as many as sixteen million, barring the development of medical breakthroughs to prevent, slow, or stop the disease. Alzheimer's disease is officially the sixth leading cause of death in the United States and the fifth leading cause of death for those aged sixty-five and older. However, it may cause even more deaths than official sources recognize. It kills more than prostate cancer and breast cancer combined.

Keeping the Brain Healthy

Studies indicate what meaningful steps people might take in order to maintain healthy brain function as they grow older. Interestingly, these are all activities that make for happy, healthy individuals regardless of age. One of the keys to preserving brain health is to offer up the very best "fuels" we can to prevent diseases from occurring in the first place. The typical

American diet often lacks the essential nutrition our brains need to function at its best, so positive dietary changes and some good dietary supplements are the best ways to improve brain health. Eating a healthy, balanced diet, including fruit and vegetables, and avoiding foods that produce cholesterol, is good for your brain, as well as your body in general.

Psychologists have learned that socializing and maintaining close relationships is also a key ingredient of a balanced life. The process of interacting with others is good for brain health and creates a sense of well-being and purpose. Acquiring new skills, such as playing a musical instrument or learning a foreign language, can be very stimulating and enjoyable as well.

Brain Health Facts as we Age

Seniors have higher rates of heart disease, cancer, high cholesterol, and high blood pressure than the rest of the adult population. These diseases, including neurodegenerative diseases of the brain and mental decline associated with aging, can often be prevented or controlled through healthy eating and regular physical activity. Most seniors are not as physically active as they could be. This can be due to balance problems or joint issues, but by just being physically active for at least one hour each day you can improve your heart health and are better able to control your weight than those who are not as active.

The majority of seniors are overweight or obese, which makes chronic diseases worse. Being overweight is frequently associated with excessive inflammation, which can be toxic to a healthy brain. You are never too old to lose weight, and losing just a little — even five to ten pounds — can make a huge difference in your health.

The majority of senior men and many senior women consume more than the recommended amount of sodium (salt). Decreasing the amount of sodium in your diet can substantially reduce your risk of high blood pressure. Excessive blood pressure is the greatest risk factor for stroke.

Many seniors do not get enough protein, calcium, folic acid (folate), Vitamin B6, Vitamin B12, Vitamin C, and Vitamin D through the food they eat. The right vitamins and minerals, in the right amounts, can help prevent anemia, depression, and memory loss. They can also help you heal better after surgery or an injury, and help keep your bones and teeth healthy and strong.

Most senior women do not consume enough vegetables and fruit, grains, milk and milk products, meat and meat alternatives. Most senior men do not get enough vegetables and fruit, or milk and milk products.

In 2006, a study funded by the National Institutes of Health (NIH) and the National Institute of Nursing Research (NIR) concluded that "exercising" the brain could very well lead to cognitive improvement. The study was conducted among people age sixty-five and older. Three groups of participants were given computer-based training for either memory, reasoning, or processing speed (the ability to respond to computer screen prompts). A segment of each group received reinforcement training eleven months later. A fourth group received no training whatsoever. All of the trained participants were tested before and after the initial and reinforcement training sessions. Everyone, including the untrained, was evaluated once a year for a period of five years. Further, individuals who received training reported having less difficulty with everyday activities such as preparing meals, managing finances, and handling household chores, with significant improvement among those who were given reasoning training. Researchers concluded that the training roughly counteracted the degree of decline in cognitive performance that would be expected over a seven to fourteen-year period among a comparable age group without dementia.

Omega-3 Helps Preserve Brain Size in the Elderly

The medical journal *Neurology* reported a study, led by James V. Potalla, PhD at the University of South Dakota, of women age seventy years on average. His team reported that women with higher levels of omega-3 fats in their blood at the start of a study had a 0.7 percent larger brain volume than women with lower levels eight years later. The part of the brain that plays a major role in memory formation, the hippocampus, was 2.7 percent larger. The researchers suggest in their findings that omega-3 can be helpful in slowing down age-related brain atrophy. Brain shrinkage tends to be accelerated in those with Alzheimer's disease, making preserving the brain a key facet of Alzheimer's disease prevention.

Dr. Potalla said, *"Omega-3s are building blocks for brain cell membranes... If achieving certain omega-3 levels can prevent or delay dementia, that would have huge mental health benefits, especially since levels can be safely and inexpensively raised through diet and supplementation."* Professor Potalla also stated that previous studies have shown that eating non-fried oily fish twice a week and taking fish oil supplements can raise the mean red blood cell level of EPA and DHA to 7.5 percent—the same level as the women with the highest omega-3 levels had in this study examining brain size.

Zaldy Tan, M.D. of the University of California Los Angeles, and colleagues, have noted that structural findings in the brain suggest that people with low levels of the nutrients, found mainly in fish, have brains that appear to have aged faster than normal.

Fish Oil and a "Smarter" Brain

What is the role of omega-3 in those that don't have dementia? Can fish, or the omega-3 fatty acids found in fish, make you smarter? The answer is "quite possibly!" Numerous studies have shown the cognitive benefits to the developing fetus and young children whose mothers consumed greater amounts of omega-3 during pregnancy. One of the early studies that led researchers to investigate this omega-3s "smarts advantage" came from studies looking at distribution of body fat. William Lassek of the University of Pittsburgh, and Steven Gaulin, of the University of California, Santa Barbara, reported that women with a tendency to have more fat on their hips than around their waists had higher cognitive test scores, as did their children. The researchers theorized that because fat on hips is higher in omega-3s than belly fat, these women were storing omega-3s important to fetal and infant brain development, as well as boosting their own brainpower.

In a study done in 2014 researchers reported that eating fish and providing more omega-3 could have an effect on the development of human fetuses. They looked at the data from 151,880 mother and child pairs. They found that women who ate fish *more* than once a week had lower risk of preterm birth than women who ate fish only once or fewer times a week. The higher intake group also delivered babies with a slightly higher birth weight, which were therefore most viable.

A more recent study indicated that girls who had more omega-3 fatty acids in their diet outsmart those who consume higher amounts of omega-6 fatty acids. To test their assumption, Lassek and Gaulin analyzed data on about four thousand girls and boys between the ages of six and sixteen who were part of a study to assess health and nutritional status of children and adults. After considering factors such as parents' income and education, the children's age, race, number of siblings, and blood lead levels, the researchers found that girls who ate more omega-3 scored significantly better on four cognitive tests, including an IQ test.

Boys also perform a bit better on cognitive tests if they ate more omega-3s than other fatty acids, but the effect is "twice as great in girls as in boys", Lassek reported. Lassek and Gaulin also found that excessive omega-6 fatty acids *interfere* with cognition, because girls who ate more of these oils didn't perform as well.

Other researchers who have studied omega-3s are not surprised to see the link between omega-3s and intelligence: "Deficiency in omega-3 intake in modern diets is associated with an increased risk of violence, major depression, suicide, and bipolar disorder", said Dr. Hibbeln. Finding a connection to another aspect of brain function makes sense, particularly because neurons use fatty acids to build brain cells. According to Dr. Hibbeln, the omega-6 finding is most important, saying ... "the big change in the Western diet of the past one hundred years is a massive increase in the dietary intake of omega-6 fatty acids."

Another group of University of Pittsburgh researchers were interested in what could be done to enhance brain function in young adults. The team investigated a group of eighteen to twenty-five year olds, and found that youngsters who took omega-3 fish oil daily for six months had significantly improved working memory as compared to those who took a placebo. Working memory is what we use for last minute studying before a big test. It is also the type of memory used to remember a phone number that someone just shouted to you from across a room. As you may guess, it is generally our weakest type of memory, but is it still critical for normal learning and helps us focus on a goal, avoid distractions, and even resist impulsive, potentially dangerous choices. This study is profound because it was the first to confirm prior studies with older adults that found that fish oil supplements could also improve various aspects of memory in younger people.

In order to insure that these young people were actually taking fish oil during the study, researchers checked blood levels for the molecules EPA and DHA at the beginning and end of the six-month study. These molecules have important brain function. In fact, thirty to forty percent of our brain cell's membranes are made from DHA. Those with the lowest levels of DHA in the blood prior to taking fish oil were the same group to show the greatest improvements in working memory.

Are there brain benefits from omega-3 in middle age? A study done in 2004 on 1,613 people, ranging from forty-five to seventy years old, reported that those that either took fish oil supplements or ate large amounts of fish were less likely to have impaired cognitive function and speed over the five-year testing period. Interestingly, they also found that subjects who had a higher dietary cholesterol intake were significantly associated with an increased risk of impaired memory and mental flexibility.

With few exceptions, most people begin to notice some memory lapses after the age of fifty. Fortunately, for most these are manageable and do not profoundly affect our typical daily activities. However, as we get closer to our seventies and eighties, problems with thinking and memory can be overwhelming. The fact is that people over eighty-five years of age have a fifty percent likelihood of developing clinical dementia.

With this in mind, there is some recent good news about interventions you can start today — which for most people can be as easy as going to the store — that may significantly help improve your memory. Researchers in Australia recently reported on a study of women ranging in age from sixty-four to seventy-nine years who had significant improvement of memory retrieval after four months of taking a multivitamin containing a wide range of ingredients. They found that those taking supplements had significantly lower levels of homocysteine (an inflammatory molecule) associated with dementia. The multivitamin contained B Vitamins, known to reduce homocysteine levels (that can also contribute to the production of proteins), DHA (an omega-3) and myelin (a part of brain cells), in addition to the neurotransmitters dopamine, norepinephrine, and serotonin.

Fish and Expectant Mothers

The high level of omega-3s along with protein, selenium, iodine, and Vitamin D, are the primary reason why doctors and nutritionists recommend consumption of fish. In fact, some doctors say the need for omega-3 fatty acids is critical during pregnancy because they are building blocks for the fetal brain and retina. A study among animals deprived of omega-3s showed that offspring were born with visual and behavioral deficiencies that could not be reversed, even with supplementation. It is generally recommended that pregnant women should get an average of 200 mg/day of the DHA component in omega-3 as a supplement.

Summary

It is not a surprise that fish is often called brain food. Omega-3 found in fish and seafood is essential to both the structure and function of brain cells. There is a clear link between consuming adequate amounts of omega-3 and having reduced risk for conditions such as dementia, depression, attention-deficit disorder, dyslexia, schizophrenia, and even Alzheimer's disease. Omega-3s are critical for fetal and child brain development, and dietary supplements of omega-3 may be superior to eating fish due to concerns about levels of toxins in seafood during this time period.

CHAPTER

6

Emotional and Mental Health Benefits from Omega-3

The World Health Organization (WHO) and Centers for Disease Control (CDC) share the same definition of mental health as "a state of well-being in which every individual realizes his or her own potential, can cope with the normal stresses of life, can work productively and fruitfully, and is able to make a contribution to her or his community." Life circumstances and genetics play a significant role in our ability to maintain a sense of well-being. While many of us are capable of working through the stresses of life, others among us are not. Fortunately, there are a number of therapies known to have various benefits for managing mental disorders. They include alternatives as diverse as psychotherapy, counseling, meditation, breathing exercises, forms of yoga, and prescription medications. All have their place, but the latest research has shown that diet and proper nutrition can also play a role. In this chapter we will explore some of the newest information on omega-3 use for a variety of emotional and mental health benefits.

Depression

Clinical depression is characterized by the persistent incidence of sadness, inappropriate crying, and feelings of worthlessness, hopelessness, or helplessness. It's not unusual for us to experience any of these feelings at some point in our lives, but if they occur persistently, investigation by a medical professional may be warranted. Depression affects approximately

14.8 million American adults, or about 6.7 percent of the population of the United States, age eighteen and older, in a given year. Symptoms of depression may also indicate some type of underlying physiological disorder. Various physical symptoms including fatigue, lethargy, and weakness could be signs that something else is going on such as chronic fatigue syndrome, diabetes, fibromyalgia, or Parkinson's disease.

Link Between Depression and Eating Fish

This study done in 1998 and repeated numerous times since strongly links lower consumption of fish (and the omega-3 fish provides the brain) to a great risk of depression.

The Lancet, Volume 351, Issue 9110, 18 April 1998, Pages 1213

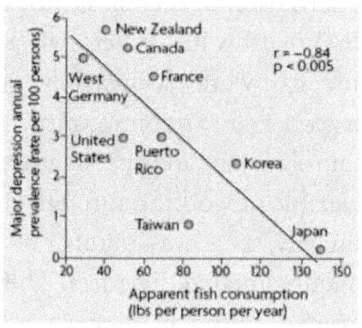

Several research studies have shown that symptoms of depression are more common in people who consume lesser amounts of fish or dietary omega-3s, which has led to a new area of potential treatments for depression. In 2001, a nationwide study of Finnish adults found that people who consumed fish infrequently were significantly more likely to have symptoms of depression than those who ate fish more frequently. Another study revealed that depressed patients have depleted levels of omega-3 EFAs (particularly of DHA) in their blood plasma and cell membranes. Currently, the American Psychiatric Association now classifies omega-3s as a complementary treatment for depression.

There is now evidence that major depression is accompanied by production of inflammatory molecules known as cytokines. A dietary imbalance, with greater amounts of omega-6 (a source of inflammatory molecules in our diet) compared to less omega-3 fatty acids can also result in an overproduction of inflammatory cytokines. A number of studies have shown that by administering just EPA (the stronger anti-inflammatory component of omega-3s) to depressed patients can bring the inflammatory state into balance and help mitigate symptoms. In 2009 an article published in the American Journal of Clinical Nutrition evaluated twenty-nine studies and reported not only a benefit of omega-3 fats in people with depressive illness, but also that the greatest benefits were in individuals with more severe depression.

Postpartum depression is a potentially serious and rapidly developing type of depression. A study in 2002 noted that the lower the content of DHA in breast milk, an indication of the mother's depleted omega-3 level, the greater the likelihood of the mother developing postpartum depression; just another reason prenatal omega-3 supplements can be essential for both mother and child.

Depression is a very common side effect after a person suffers a stroke. This is mostly due to the loss of function. A 2014 study revealed a novel ability of omega-3 to support long-term protection against narrowing of blood vessels, which supply oxygen to the brain. Further, the researchers found that omega-3 was able to stimulate existing blood vessels to grow new ones after stroke, thereby increasing blood flow to the affected area. Their findings indicate that supplementation with omega-3 has the potential to enhance tissue repair within the brain and improve long-term functional recovery after stroke and potentially reducing post-stroke depression risk.

Bipolar Disorder

Bipolar disorder, sometimes referred to as manic-depressive illness, is a psychiatric condition that may be associated with extreme disability and early death. It is less common than depression, but is also believed to have an inflammatory component. Supplementation with omega-3s has been shown to assist in the treatment of bipolar disorder.

In 1999 A. L. Stoll and colleagues conducted a double blind, placebo-controlled trial involving thirty patients with bipolar disorder and found that omega-3 supplements along with standard treatment had a significant beneficial effect in mitigating symptoms.

Schizophrenia

Schizophrenia is one of the most disabling of all psychiatric illnesses. Pharmacologic treatments for the disease often leave schizophrenic patients with residual symptoms, lethargy, and cognitive impairment.

According to D. F. Horrobin, a renowned expert in schizophrenia, decreased levels of DHA and EPA can impair the normal function of neurotransmitters, the chemicals that are used within the synapse that allow the brain cells to communicate. This suggests that omega-3 supplementation has the potential to normalize neurotransmitter receptors.

Various studies have shown benefits of omega-3 supplementation in patients with chronic schizophrenia who were not being treated concurrently with conventional antipsychotic medications. Other studies have indicated a reduction in the severity of symptoms ranging from seventeen to eighty-five percent when one to two grams per day of omega-3 were added to patients' usual medications. Similar studies with schizophrenic patients on antipsychotic medications have not shown similar benefits with supplementation of three grams per day of omega-3.

> It's important to note that despite the positive effects omega-3s might have on mental disorders, they are not a substitute for medications that a physician may have prescribed. Do not discontinue any medication without first consulting your physician.

CHAPTER

7

Fish Oil and Joint Health

At almost any age seemingly minor joint injuries from sports, falls or just twisting can result in micro-tears and other structural damage that can accumulate over time. Eventually minor injuries can lead to chronic inflammation and pain if the joint is not allowed to heal completely. The inflammation of the joint then can become chronic and eventually lead to that dreaded diagnosis of – *arthritis!*

Increasing human life expectancy means that chronic diseases like arthritis, heart problems, dementia, and diabetes are going to affect more and more people. About one in five Americans currently suffer from arthritis. Many experts in disease prevention agree that maintaining an active lifestyle and preserving joint mobility can reduce the severity, or even completely eliminate, arthritic joint pain and other chronic diseases.

Support for this theory was shown in a study published in 2012 where researchers found that those who had an active lifestyle and walked about four miles throughout the day had an almost thirty percent less risk of developing diabetes compared to those who walked about half that much. By maintaining your joints and mobility you have an opportunity help reduce your risk of diabetes. Other similar studies of heart disease, and even cancer risk, show that by being active you can help maintain wellness as you age.

We have discussed the role of fish oil and omega-3s as a potent anti-inflammatory, but only briefly mentioned the danger of most prescription and over-the-counter medications. It's estimated that over seventy-five million Americans live in chronic pain. For the majority, this serious discomfort is related to joint pain. Whether from a disease condition such as osteoarthritis or rheumatoid arthritis, or more commonly from joint degeneration or overuse, Americans often turn to non-steroidal (NSAIDs) pain relievers for help. Granted these medications, which are now found in every drugstore, supermarket, and corner store in the country, can be a miracle cure for many. However, like many things that are seen as too good to be true, non-steroidal medications have an ominous potential to also cause severe pain, side effects, and even death.

You know many of their common names, ibuprofen, naproxen, and aspirin, which are available over the counter. And the prescription ones like, Celecoxib, Indomethacin and Piroxicam, to name but a few, are some of the most commonly prescribed drugs in the world. Used for several days to a week or two, these medications generally have a good safety record and are very effective at relieving joint pain and discomfort. But what can occur with long-term use can be frightening. Because most joint conditions are chronic, meaning they will continue to occur and be painful, NSAIDs are used for a much longer period of time than is typically recommended by the manufacturer.

With longer-term use, up to fifty percent of NSAID users will develop heartburn or other types of stomach pain. Almost all who take the older generation of NSAIDs, such as ibuprofen, will have evidence of stomach bleeding, and eight to twenty percent or more will experience ulceration. In addition, three percent of those on NSAIDs will develop serious gastrointestinal side effects, which result annually in more than 100,000 hospitalizations, an estimated 16,500 deaths, and a yearly cost of more than $1.5 billion to treat their complications. Indeed, NSAIDs are the most common cause of drug-related sickness and death reported to the FDA and other regulatory agencies around the world.

For these reasons, a natural approach with fewer side effects to address joint pain and stiffness should be considered. Dozens of clinical studies have revealed the benefits of fish oil supplementation in alleviating the symptoms of joint inflammation. Studies have shown omega-3 can reduce levels of inflammatory markers called cytokines in joint fluid by up to ninety percent. Fish oil supplementation can decrease the number of tender joints, the duration of morning stiffness, and overall pain in patients with conditions like rheumatoid arthritis.

Fish oil supplementation has been studied and found to relieve joint pain so efficiently that many people can stop or reduce their need for NSAIDs. For example, in 1993 C. S. Lau and colleagues compared the supplementation of 2.8 grams of fish oil supplements used daily versus a placebo in sixty-four patients with stable rheumatoid arthritis. In three months, the fish oil group showed significant reduction of NSAID use as compared with the placebo group not taking additional fish oil.

A Spine Surgeon's View

In our own neurosurgical practice we see the ravages of overuse joint pain and degenerative joint pain of the spine. Excessive weight, poor conditioning, the aging process, and improper body mechanics can all play a role, but no matter what the cause, inflammation is the root cause of their pain. Almost all of the patients we see are on high dosages of NSAIDs for a long time for their pain relief. As part of our evaluation we counsel them about the side effects of long-term NSAID use and suggest trying natural alternatives. We recently surveyed over two hundred and fifty of my spine pain patients who started using omega-3 fish oil supplements. Almost two-thirds were able to stop taking their NSAID medication, and were only using fish oil for pain relief! Not surprisingly, many commented on how much better their stomachs felt and that they were glad to be off their drugs.

Additionally, along with fish oil (omega-3), we often recommend glucosamine/chondroitin for joint pain relief along with other dietary supplements and herbal treatments. Turmeric (curcuma longa) is a fragrant yellow spice found in curry and used for centuries in ayurvedic medicine (which originated in India more than three thousand years ago) with similar anti-inflammatory effects as aspirin, but without aspirin's increased bleeding potential. Boswella is another natural herb that can block the inflammatory pathway. Bromelain is an enzyme found in pineapples that interferes with inflammatory cytokine production. White willow bark has been known for centuries as the natural precursor to aspirin, but without the stomach side effects, and can be helpful for acute short-term pain relief. Other natural products that are commonly used include green tea which is a powerful polyphenol, and antioxidant nettle leaf that has been shown to reduce immune system inflammation, and s-adenosylmethionine (SAMe) that can protect joint lining cells by reversing antioxidant depletion within the joint sac.

Better joint health is also about lifestyle and diet choices and starts with weight control, avoiding overuse injuries, and maintaining proper muscle strength and tone to support the joints. But despite even the best intentions, joints can become injured and degenerate, and arthritis can occur as we age.

Curbing the Inflammatory Response

We have already learned a lot about the body's inflammatory response, but perhaps you were unaware that it is part of the immune system. Once the immune system is activated it starts the inflammatory process like a general would call his troops into battle. Not only can these soldiers act for our benefit, but also our detriment if they remain active for longer than needed. Beneficial effects include fighting off infectious bacteria and viruses, and stimulating the healing process. However, the bad effects are some of the greatest sources of pain and suffering we endure.

Chronic inflammation can be the result of conditions known as autoimmune diseases. Autoimmune diseases are an example of our immune system behaving badly. There are well over one hundred and fifty known autoimmune diseases and conditions that can affect almost every organ in the body, but are especially common in connective tissues and joints. Autoimmunity occurs when certain cells in our body lose their unique cellular signature, and the immune system, by activating the inflammatory response, sees those cells as foreign invaders and begins to attack them. When this occurs the inflammatory response causes tissue to swell due to additional blood and white blood cells being called into the area of attack.

This leads to pain, redness, and an increase in temperature within the cellular battle zone.

Rheumatoid arthritis (RA) is one of the most common examples of autoimmune disease and affects an estimated 1.3 million Americas, most of which are women. This is a devastating disease that can lead to profound deformity and disability of joints throughout the body. Based on its severity, life-long immunosuppressant drugs such as steroids, and even organ anti-rejection drugs, are used to help curb the inflammatory response. These drugs, along with powerful non-steroidal anti-inflammatories (NSAIDs) can, however, have profound side effects and leave the immune system so weak that other diseases like cancer can be more common.

Degenerative Joint Disease

Every time we move or even breathe we are moving joints throughout our body. As we age and do physical work our joints tend to suffer from wear-and-tear and can degenerate over time. Over the last one hundred and fifty years, the average American's life expectancy has almost doubled from about forty years to nearly eighty years. Therefore it is not surprising that almost thirty million American's have been diagnosed with degenerative joint disease, also called osteoarthritis (OA). Millions more likely suffer from OA, but self-medicate the pain associated with this condition. OA can, similarly to RA, lead to profound disability and reduced mobility, and has been associated with the obesity epidemic in the United States and other countries with rapidly aging populations.

Joint degeneration found with OA also has an inflammatory component. OA is a form of injury that activates our immune response to clear injured tissue and start the healing process. The cycle of joint overuse, followed by pain, then remission is directly related to inflammation. Unlike RA where the immune system will continue its cellular attack and therefore requires complete blockade of the immune response, OA treatments are most effective given periodically when there is an inflammatory flare up, or daily at lower doses to allow the immune healing response to occur.

How to Keep Joints Healthy:

- Maintain ideal weight (BMI) – Ideal body weight will reduce pressure on joints, surrounding ligaments and muscles and improve mobility.
- Improved ergonomics – To reduce joint strain with repetitive movements by placing often used objects and machines in ways to support joints and natural joint movements.
- Proper body mechanics – Avoid back, hip and knee strain and injury by holding objects close to the body when bending and lifting. Use the bigger leg muscles to lift and avoid back twisting and pulling.
- Proper posture – Good posture means we place our vertical weight in line with gravity when sitting or standing. This puts less strain on our skeletal system and less strain on joints.
- Build muscle – By engaging in weight training and resistance exercises not only can you build the muscles that surround joints, you can strengthen bones that make up the joints. In addition, improved muscle function will improve balance and reduce that risk of falls that can injury joints and bones.
- Avoid overuse injuries – Joint injury from ignoring pain is very common in both sports and the work place. We often see this with cross-country runners. The body's feedback mechanisms activate when either a new injury has occurred or if the joint motion is beyond the scope of normal movement. In either case, to push through the pain means to you are risking further injury and chronic inflammation.
- Rest your joints – By resting joints between activities you can activate natural joint repair processes in order to prevent injury. Cool down periods followed by heat, massage, and range of motion activity all can be used to help joints recover.
- Regular exercise and range of motion – Maintaining a daily exercise and stretching program is like an annuity that will pay big dividends as you age. Start now and reap the benefits for a lifetime of better joint health.

Omega-3 for Inflammatory Joints

Since OA is generally a progressive condition, sufferers will require long-term, if not lifetime, treatment. Therefore the use of alternative or natural treatments, generally with much fewer side effects, is the best option for many. Although both over-the-counter and prescription drugs can be very effective, the trap many people fall into is using them beyond the seven to ten day maximum duration typically recommended by the drug manufacturer. Long-term use can lead to profound side effects, and even death, from conditions such as stomach bleeding and liver or kidney failure.

Dozens of clinical studies have revealed the benefits of fish oil supplementation in alleviating the inflammation symptoms of arthritis, showing that fish oil reduces levels of inflammatory pain producing cytokines in the joint fluid by up to ninety percent. In 1995 J. M. Kremer and colleagues conducted a double blind, placebo-controlled study that showed that fish oil supplementation of 1.3 grams per day decreased the number of tender joints, the duration of morning stiffness, and overall pain in patients with arthritis. In 1993 C. S. Lau and colleagues compared the supplementation of 2.8 grams of fish oil daily versus a placebo in sixty-four patients with stable rheumatoid arthritis. In three months the fish oil group showed significant reduction of prescription anti-inflammatories use as compared with the placebo group, and that reduction peaked at twelve months.

Back Talk

The most common reason people over age forty come to our office is arthritis-related pain in their back or neck. This type of arthritis, also a form of osteoarthritis, is universally found to some degree in the joints of those over age forty. Since each segment of the spine is also connected with joints, there are numerous areas of our spine that can be afflicted by this condition. Arthritic pain can have significant impact on the sufferer's quality of life. Patients with arthritis are not able to be as active as they once were. Participation in sports and other activities requiring ease and freedom of movement becomes limited. Fortunately, relief can be found for the majority of patients without resorting to surgery.

In our neurosurgical practice, most patients with spine complaints leave the office with a prescription for non-invasive treatments rather than surgery. Examples of non-invasive treatments are physiotherapy, traction, massage, ultrasound, injections, and chiropractic therapy — all designed to relieve pain, recondition the body, and restore posture and quality of life. The good news is that once the acute inflammatory phase is reduced, symptoms can become tolerable and surgery less likely to be required. This is where fish oil enters the treatment plan.

Why Long-Term NSAIDS are Bad News – Joseph Maroon MD, FACS

In 2004, two key events occurred that would forever change the way I treat my patient's non-surgical pain and what I prescribe for relief. The first event was rather personal. As an Ironman triathlete, I grew accustomed to enduring pain. At one point I began to experience severe left knee pain and attempted to counter the pain with over-the-counter (non-prescription) NSAIDs. Over roughly a two-month period, my knee began to feel better, but I developed stomach pains. These pains were diagnosed as gastritis and early gastric erosion – most likely caused by the NSAIDs. I became concerned, not only about my own well-being, but also about the thousands of patients who I had placed on these medications.

The second event had to do with a new group of prescription NSAIDs that came on the market that were designed to cause fewer stomach problems. These prescription NSAIDs therefore seemed like a better alternative for my patients and me. However, this soon proved to be an illusion. During a clinical trial, one of these new drugs, Vioxx®, was found to be associated with an increased occurrence of serious cardiovascular events, including myocardial infarct and strokes. On September 30, 2004, the FDA acknowledged the voluntary withdrawal of Vioxx®—a decision that ensured that the use of COX-2 NSAIDs would never again be considered standard care.

I later learned from a friend about omega-3 fish oil for joint related inflammation, and I personally began to take two grams of omega-3 fish oil supplements every day. After two months, I had no further knee pain, was able to train again and, most notably, I had no further stomach problems.

After essentially banning NSAIDs from our practice, we soon found several hundred peer-reviewed articles in scientific medical journals that reported on the use of omega-3s for a host of inflammation-related joint conditions. The omega-3s appeared to block the inflammatory response in a similar, but safer, way than other anti-inflammatory drugs.

In 2005, we conducted our first research study on fish oil supplements. We surveyed two hundred and fifty non-surgical spine pain patients for whom we had also recommended omega-3 fish oil supplements. The survey showed a surprising number—sixty percent—had obtained significant spine pain relief and almost the same amount—fifty-nine percent—were able to stoptaking NSAID pain medications and rely solely upon omega-3 supplements for pain relief (Graph 1). The results of the trial mirrored other controlled studies finding that the NSAID ibuprofen and omega-3s were equivalent in reducing arthritic pain. Omega-3 fish oil supplements appear to be a safer alternative to NSAIDs for treatment of nonsurgical neck or back pain in this selective group. The results of our trial mirrored other controlled studies finding that the NSAID ibuprofen and omega-3s were equivalent in reducing arthritic pain. Omega-3 fish oil supplements appear to be a safer alternative to NSAIDs for treatment of nonsurgical neck or back pain in this selective group.

The results were published in *Surgical Neurology International,* which prompted an editorial from the senior editor to encourage more neurosurgeons to learn about alternative treatments.

Graphs– Results of a survey of one hundred and twenty-five respondents on omega-3 fish oil for a minimum of two-months that had been seen with non-surgical spine pain.

Graph 1. Shows that the majority of patients their NSAIDs and pain meds after two months on fish oil.

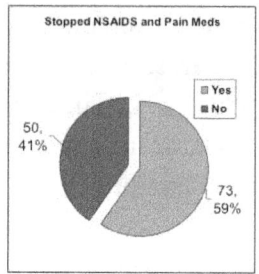

Graph 2: The majority had less pain symptoms when using fish oil.

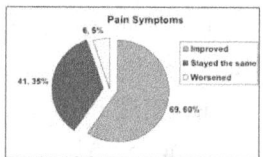

Graph 3: Joint pain improved in 60 of the stopped respondents.

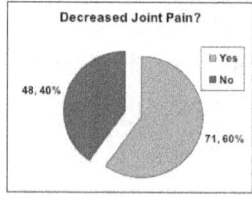

Graph 4: Eighty percent were satisfied with their experience with fish oil.

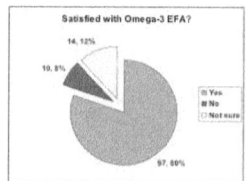

We continued to use omega-3 supplements as the first line of treatment for patients with non-surgical spine pain, and continue to have tremendously positive results. As with any supplement, we urge patients to take a high quality pharmaceutical grade product. To date, the vast majority of patients are very happy and highly compliant with these supplements and report almost no side effects.

The recommended dosage range for most inflammatory conditions is of 1.5 to 5 g of combined EPA and DHA per day, taken with meals. (See prior section on ideal ratios.) For the study conducted with our patients, the dosage ranged from 1.2 to 2.4 grams per day. Since that time, tests have been developed to determine omega-3 blood levels to more accurately dose an ideal level for any given patient. In general, an initial higher loading dose of fish oil may be needed if acute pain is present.

Potential Side Effects

Natural ways for pain reduction resulting from inflammation are generally much safer than NSAIDs and can be taken daily for as long as needed for most people. However, fish oil is known to alter blood-clotting capacity, typically a benefit for people with a risk for heart disease and stroke, but possibly undesired if a person is taking blood-thinning medications. Use of fish oil as a dietary supplement should be approved by your healthcare professional if you are on prescription medications, and more importantly if you are on medications like Coumadin° or other blood thinners. Those who are allergic to fish or seafood should not take fish oil supplements due to the risk of an adverse side effect.

Coumadin® is a registered trademark of Bristol-Myers-Squibb

CHAPTER

8

Fish Oil for the Rest of the Body

Omega-3s and Eye Health

We have mentioned several times the importance of the omega-3 DHA for brain health. The major reason for this is the structural cell membrane of nerve cells are mostly made of DHA. For brain cell membranes, over forty percent is DHA and for the retina of the eye, over sixty percent is DHA. Since DHA is an essential fatty acid and the body cannot make omega-3s it is critical we consume enough in our diet, especially for those areas of our body that have higher concentrations of nerve cells. In addition to the brain and retina, high concentrations of nerve cells include those that help to control heart rhythm, cells critical for normal spinal cord, peripheral nerves in the arms and legs, and cells to help make skin watertight (as we will learn later).

Fetal and infant eye development is critically dependent on adequate amounts of maternal omega-3. Researchers have shown that infants whose mothers received DHA supplements from their fourth month of pregnancy until delivery were less likely to have below-average visual acuity at two months of age than infants whose mothers did not receive the omega-3 supplements. And numerous studies have found that healthy pre-term infants who were fed DHA-supplemented formula showed significantly better visual acuity at two and four months of age, compared with similar pre-term infants who were fed formula that did not contain the omega-3 supplement.

As a neurosurgeon and a physician assistant in neurosurgery, we often evaluate and treat diseases of the eye. Although not all eye conditions can be helped with dietary supplements, there is a role for using several natural dietary supplements. Historically, various combinations of vitamins, minerals, and dietary supplements have been used to not only enhance general eye health but also to treat or attempt to prevent the development of common eye conditions such as dry eye, cataract, glaucoma, macular degeneration, and diabetic retinopathy. Among the earliest used and still the most common are the antioxidants: Vitamin C, E, beta-carotene, carotenoids (Lutein & Zeaxanthin) and omega-3 essential fatty acids.

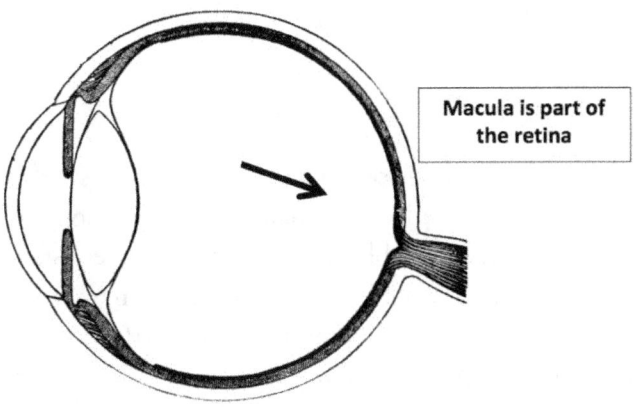

Macula is part of the retina

Retina – The Nerve Center of the Eye

The retina is a thin layer of tissue that lines the back of the eye on the inside. It is located near the optic nerve. The purpose of the retina is to receive light that the lens has focused, convert the light into neural signals, and send these signals on to the brain. The macula is part of the retina that is where most of the nerve receptors for both our central fine vision and color vision are located.

Omega-3 for Eye Disease

Omega-3 fish oil has numerous properties that can affect the development of unwanted new blood vessels in the retina and improve cell membrane repair and speed of signal transmission critical to normal eye function. In addition to maintaining good eye function, DHA in particular, can protect against macular degeneration. Specialized molecules in the retina have very high levels of DHA. These molecules are very sensitive to reduced blood flow, excessive light exposure and oxygen/energy metabolism changes. Adequate omega-3s in your diet can help prevent retinal cell death, oxidative stress, reduce age-related inflammation, and developmental processes associated with the aging eye.

Advanced Macular Degeneration

An estimated eight million persons at least fifty-five years old in the United States have intermediate or advanced macular degeneration (AMD.) AMD is the most common cause of blindness for people over age sixty. Of the eight million, 1.3 million would develop advanced macular degeneration if no treatments were given to reduce their risk. AMD is the leading cause of vision loss in people aged sixty-five years or more. Early forms of AMD are characterized by deposits of extracellular material in the retina. These deposits are known as "drusen." Drusen are protein–carbohydrate–lipid complexes present in most people diagnosed with AMD.

Odds of AMD with visual loss have been shown in numerous studies to be less with increasing intake of DHA. Several studies that looked at the relationship of fish intake with 'late' Age-Related Maculopathy (ARM) also show somewhat slower progression with a greater intake of fish.

In 2013, the National Eye Institute (NEI) published results of a large study that investigated whether daily supplementation with omega-3s, along with nutritional a supplement containing beta-carotene, Vitamin C, Vitamin E, zinc and copper would further reduce the risk of AMD progression among study participants with early signs of macular degeneration. By adding omega-3 they reduced the risk of AMD progression by an additional twenty-five percent among a similar population.

Diabetic Retinopathy

Diabetic retinopathy is the most common retinal disease among residents of the United States aged eighteen to fifty-five. That equates to approximately 5.3 million individuals, with some twenty-four thousand people becoming blind each year by the disease. Diabetic retinopathy is due to both excessive new blood vessel formation and neural degeneration. Omega-3s, both EPA and DHA, have been shown to reduce associated inflammation, excessive blood vessel growth, and blood clotting from diabetic retinopathy.

The fish oils, EPA and DHA, also appear to have myriad effects on the eye. DHA, when combined with Vitamin A, has been shown to slow progression of retinitis pigmentosa (RP), an inherited, degenerative eye disease that causes severe "tunnel vision" impairment in certain patients.

Dry Eye Disease

Dry eye disease (DES) is prevalent and increasing in incidence, especially with advancing age. Dry eye is a serious detriment to a patient's quality of life, and many millions are spent on over-the-counter lubricants and drops. Inflammation plays a significant role in DES. For example, increased concentrations of cytokines (inflammatory markers) have been found in the tear film of dry eye patients.

From a prospective, randomized study in 2013, researchers evaluated two hundred and sixty-four patients with dry eye. The treatment group received 650mg EPA and 350 mg DHA daily for three months. The results showed sixty-five percent improvement in the treatment group. There was a significant eye examination changes in the omega-3 group.

Skin Health and Omega-3

Our body contains up to seventy-five percent water, if it was not for our waterproof skin layer we would never be able to survive. Although our skin is mostly waterproof, the cells that make up the deeper layers also need proper hydration. In addition to maintaining proper hydration, a balanced diet is critical to skin cells functioning properly. Omega-3s are critical to the structure and function of skin cells.

Omega-3s need to be replaced in our diet frequently due to rapid skin cell breakdown. Skin cell oxidative damage and death occurs from environmental factors such as UV sunlight, smoking and other pollutants, as well as from an unbalanced diet (excessively inflammatory), lack of exercise, stress, and illness. Oxidative production of free radicals can lead to accelerated skin aging. Free radicals promote oxidation of skin cell DNA, proteins and lipids, including omega-3s, which damage intracellular structures.

Omega-3 fatty acids bolster the skin cell membrane of the epidermis layer of skin. By doing this, omega-3 fatty acids in the membrane can improve cell function to improve the ability of the cell to hold onto water. Therefore, if the skin cell holds onto water, it leads to moister, softer skin, which promotes wrinkle prevention. Omega-3 fatty acids contribute to the upkeep of the skin cell membrane, improving the texture and quality of skin.

Inflammation of the Colon

Ulcerative colitis (UC) and Crohn's disease (CD) are chronic idiopathic (i.e., the cause is unknown) inflammatory disorders of the gastrointestinal tract, collectively termed as Inflammatory Bowel Diseases (IBD). Crohn's disease (CD) is a condition of the intestinal tract that can present at any age and result in significant abdominal pain and unwanted lifestyle changes. It is characterized by chronic intestinal inflammation that can be chronic or with periods of exacerbation and remission often related to stress or some other environmental trigger. New research shows intestinal bacteria play a role, and natural treatments that include both probiotics and omega-3 have shown to help. Omega-3 has numerous actions to reduce inflammation, such as seen in the bowel of Crohn's disease patients.

Randomized placebo-controlled studies that evaluated the effect of daily intake of capsules containing omega-3 to maintain remission in Crohn's disease were recently reviewed in 2014. The six studies of one thousand thirty-nine patients found a benefit for omega 3 fatty acids over placebo in preventing relapse of disease at one-year period. There were no serious side effects in any of the studies.

Ulcerative colitis can generally involve a larger area of the intestinal tract and may present with more severe symptoms. Studies investigating the benefits of omega-3 have been mixed, but a significant benefit with fish oil was found in the subgroup of patients at high risk of relapse resulting in fewer recurrences over the duration of the study.

CHAPTER

9

Understanding Omega-3 Research: What to Believe?

We are bombarded daily by the media with information overload. Eventually many tune out and focus only on what directly affects our own lives. Those in the media know this and often promote health related news as "critically important" or "breaking news". Sometimes the information is important and action is needed but often the sensational headlines lead to wrong conclusions and even fear that can be harmful. The most impactful stories the "_____scare". Just fill in the blank: Tylenol® scare, Swine Flu scare, contaminated meat scare, Zike virus scare. Each one draws instant attention and interest from the now "scared" public. Scientific journals also know that "scare" stories will be covered by the media, and that some obscure journal for the lay public will make the news.

You Be The Judge: The Facts behind Prostate Cancer and Omega-3 Fatty Acids

In July 2013 a study was published that caused a media frenzy. The headlines screamed: *"Men who take fish oil omega-3 supplements at seventy-one percent higher risk of prostate cancer"* and *"Fish oils may raise prostate cancer risks, study confirms."* Media reports went on to report the author's conclusions that *"fish oil supplements and other nutritional supplements are dangerous and harmful"*.

Not only was this a poor study there are dozens of other studies—even one from this study's primary author—showing omega-3 fish oil is *actually beneficial for prostate health*. Once scientists from around the world began to dissect the study they found major errors that made it hard to believe it was even published. First, the investigators used unrelated data from an older study by another group of scientists, not even investigating omega-3s, which had started over ten years prior. It was a study to determine the possible benefits of selenium and Vitamin E to prevent cancer. Of the over thirty-five thousand men in this other study, which started in 2001 and ended in 2007, eight hundred and thirty-four were diagnosed with prostate cancer. Those with cancer were then subdivided into low-grade (684) or high-grade cancers (156). Of the original thirty-five thousand five hundred men (less than 0.004 percent), were found to have the high-grade prostate cancer. As part of this older study, one blood sample was obtained on these men *after their cancer diagnosis* in order to test for various types of fatty acids, like omega-3s, in their blood.

Using the results of this one blood test the authors found that trans fats (which are also know to contribute to blood vessel disease and *is __not__ fish oil*) were higher in those found to have cancer. *Additionally they found DHA and EPA, which are found in fish oil, were not significantly higher in those with cancer.* What they did find was the total omega-3, which is found in flax seed, walnuts, most types of fish and seafood, grass-fed beef and tofu, was higher in those that now had prostate cancer.

Despite this being only one blood test and with no records of actually how much and of what source these men obtained omega-3, the media headlines singled out fish oil as the cause of this cancer. As one critic to the study asked, *"How do we know that prostate cancer itself is not responsible for the elevation in omega-3"? "Doesn't cancer kill cells and release the cell membranes into the blood and aren't cell membranes made of omega-3"?*

Now you be the judge. Was the media right to print these headlines? Were the author's conclusions valid? Was the media right to single out fish oil and yet no record of fish oil use as a supplement was recorded? Are there other facts about this study that are being missed and should they have been reported? Should we ignore decades of studies indicating omega-3 and prostate cancer prevention and the other health benefits of omega-3 fish oil? Read theses "facts" below and you decide.

Facts – Not Reported by the Media

1. The study did not report, nor did they ask their subjects, whether or not they consumed fish oil supplements, fish, seafood, or any other type of omega-3 fatty acids in their diets.
2. The study concluded that omega-3 fatty acids enhanced prostate tumorigenesis (tumor cell formation) and yet did not show that prostate cancer was in fact caused by omega-3.
3. The study found that trans fatty acids (TFA) were higher in prostate cancer subjects, yet despite this, it concluded that the total omega-3 caused an increased cancer risk even though there was no documentation of the subjects ingesting omega-3s or fish oil.
4. The media never reported that the study had found that both low and high grade prostate cancer groups also had had a higher prostate-specific (PSA) score (which is a marker for existing prostate cancer), nor that a greater proportion of first-degree relatives with a history of prostate cancer prostate cancer ended up with cancer compared with non-cancer controls.
5. The same study authors published a study in 2011 that looked at a larger number of prostate cancers and fatty acid blood levels using data from the Prostate Cancer Prevention Trial and found no relationship to total omega-3 blood levels and the one thousand eight hundred and nine reported cancers found in this study.
6. The study authors based all of their results on only one blood test done at the time the prostate cancer was diagnosed. Consider the following:
   a. Does cancer itself alter dietary fatty acid levels? – YES
   b. Does a single meal of fish significantly raise total omega-3 levels before a blood test? – YES
   c. Does your omega-3 level change week-to-week and month-to-month based on your diet? – YES

7. The levels of omega-3 found in this study were about forty percent of what would be expected for a person who would normally take fish oil supplements. Therefore:

    a. The author's statement that "fish oil supplements and other nutritional supplements are dangerous and harmful" is most likely **FALSE** since the subjects were most likely not taking fish oils supplements.

8. If the subjects were actually not taking fish oil supplements, then what was the source of their omega-3? – Unasked and Unknown

9. Media reports generally ignored multiple studies with more recent data that have shown ***reduced rates of prostate cancer in populations with diets rich in omega-3 polyunsaturated fats.***

10. Finally, toxicity data submitted to the FDA for prescription omega-3 products for Lovaza®, now a $1.2 billion per year prescription fish oil, did not suggest mutagenic, genotoxic or carcinogenic potential.

Perhaps the most interesting facts overlooked by the media are the numerous other studies published by the lead author Theodore M. Brasky on the benefits of fish oil. In a study published in January 2013 he found that DHA and non-fried fish consumption provided a benefit in the primary prevention of pancreatic cancer. In another study evaluating post-menopausal women and the risk of breast cancer he concluded that "trans fatty acids and linolenic acid (omega-6 EFA) was positively associated with increased risk of breast cancer and intake of EPA and DHA fatty acids reduced the risk". How could omega-3 go from preventing cancer to causing it?

This was a poorly designed study. Using data from another study, not knowing their subject's fish oil use or fish consumption and drawing conclusions ***based on one blood test*** results, which can change daily, is inappropriate and misleading. Yet these results were widely presented in the media as scientific fact. Pronouncements by the media and certain researchers served only to scare the public and distract from the positive health results of thousands of other research trials that have demonstrated consistent, positive results from the use of EPA and DHA from food and supplement sources. The greatest danger following a such a study, and the subsequent media's fear mongering, would be for the public to stop taking fish oil without getting the complete story we and others have provided.

What Omega-3 Fish Can and Can Not Do

In another article published in March 2014 in The Journal of the American Medical Association (JAMA) authors who were investigating the use of vitamins and omega-3 supplements for the treatment of age-related macular degeneration, a type of eye disease, reported an unrelated finding about fish oil and heart disease. In their conclusion they reported that those subjects, average age of seventy-four, that were give a daily fish oil supplement designed for eye disease, did not reduce cardiac event (such as heart attack and congestive heart disease) *in those who already had heart disease*!

This was amazing news and the media blasted headlines everywhere, "Study Finds No CVD Benefit With Omega-3 Fatty Acids," "Studies question fatty acids' heart benefits," and "Supplements Didn't Reduce Heart Disease in Elderly". Only rarely was it reported this was an eye disease study, the dose of the fish oil was for eye health, or that only the subjects who had preexisting cardiovascular disease showed no benefits. And to make matters worse, due to the study size, they could only predict a benefit of greater than twenty-five percent. Therefore, any benefit from fish oil less than twenty-five percent could not be discovered. The public was again left with only half-truths and cynicism as to the real benefits of omega-3 supplements that the American Heart Association and many cardiologists now recommend for their patients.

This brings us back to the concept of disease prevention and maintaining wellness. Imagine you are sevety-four years old, living your whole life as a two-pack per day smoker, and you are told to take this pill every day and to keep smoking, and all your disease risks from smoking would be reduced by at least twenty-five percent or more. You would immediately recognize this as a laughable claim and realize it is impossible. Why? First you have most likely smoked for fifty years, you know you already have shortness of breath and other smoking related problems and you are going to continue to smoke, so how could any pill (or dietary supplement) stop the effects of smoking?

Whether its smoking or heart disease, taking prescription medications or dietary supplements rarely works miracles to reverse long-standing disease, especially if the underlying lifestyle factors have not changed. A smoker knows in order to improve his or her health they have to stop smoking. The same is true with most chronic health conditions. It requires a comprehensive approach, first to refrain from unhealthy lifestyle choices and then to start using the tools known to improve your health. Exercise, improved diet choices, weight control, reducing stress and toxins is the first line of prevention, followed by dietary supplements known to help reduce the risk of disease from starting in the first place. This means starting NOW! Like money earning interest in the bank, the earlier you begin to realized that your choices today will pay big dividends and affect your health for the rest of your life the better. This book is just the start of that journey to better health. Fish oil supplements are just part of a healthy living plan that starts with what you do right now. Our hope this book and the information you now have can help you make the *right choices*!

CHAPTER
10

Take This Advice to Heart

Fish Oil and Cholesterol Lowering Medication

I think it would be helpful to begin with a very brief overview of cholesterol and triglycerides and how they relate to heart and vascular disease.

- Cholesterol is a naturally occurring fat made largely in the liver. Its purpose is to build cell membranes, certain hormones, Vitamin D, and other substances important to body function. While we get some cholesterol from foods we eat, it is not the same as that which our body makes. Cholesterol from food can cause blood cholesterol levels to rise in some people, but it does not accumulate in blood vessels or cause atherosclerosis (narrowing and hardening of the arteries).

- Cholesterol cannot circulate through the body on its own. It needs to be carried by proteins such as LDL and HDL. What we call "bad" cholesterol, LDL, is low-density lipoprotein. LDL transports cholesterol from the liver to the cells. What we call "good" cholesterol, HDL, is high-density lipoprotein. HDL transports excess cholesterol from cells back to the liver to be broken down and eliminated.

- Triglycerides are another type of naturally occurring fat and are created from unused food calories and stored in fat cells. When needed for energy, hormones trigger the release of triglycerides into the bloodstream. These particles are also unable to circulate on their own, and are carried through the body by VLDL, very low-density lipoproteins. Like cholesterol, triglycerides also contribute to atherosclerosis.

- The "total cholesterol" included on blood chemistry laboratory reports is the sum of LDL, HDL, and VLDL. While lab reports include LDL and HDL separately, they typically do not show VLDL. That explains why your total cholesterol score is higher than your LDL and HDL when added together.

Cholesterol becomes far more toxic when reacting with free radicals in your body oxidizes it. Oxidized cholesterol causes an inflammatory response in the vessel wall, allowing for more cholesterol deposition. This inflammatory response can be important in the development of heart attacks. Changing the type of cholesterol in vessel walls to a non-oxidized, non-inflammatory state would be beneficial, even if medication didn't lower the blood level or remove the cholesterol from the vessel walls.

It has long been established that high levels of total blood cholesterol along with elevated levels of LDL can increase your risk of heart and vascular disease. Because heart disease remains the number one killer of Americans, for men and women, a huge portion of our healthcare dollars are spent on risk reduction for this condition. In addition to not smoking and maintaining normal blood pressure and weight, reducing cholesterol levels has been a major focus to reduce heart disease risk factors.

Doctors have long advocated a low fat diet and exercise as the first steps to reign in cholesterol, but like so many health problems, Americans often would often rather just take a pill. In 1971 scientists in Japan discovered a molecule that could block the production of cholesterol in the liver and thereby reduce the overall amount of cholesterol in the body's circulation. That discovery has led to the development of the class of cholesterol lowering drugs known as statins, which are now the most prescribed medications in the history of the world.

Even before statins and other drugs were used to lower cholesterol, a number of natural dietary supplements had been show to reduce cholesterol and triglycerides. Among the first of these were molecules known as phytosterols.

Phytosterols are derived from certain grains and seeds, and are also available in dietary supplements. They have been shown to reduce cholesterol levels and significantly impact the cause of most heart attacks and strokes. In fact the U.S. Food and Drug Administration endorsed phytosterols *as part of a dietary strategy to reduce the risk of coronary heart disease.*

Most recently research has shown that omega-3 fish oils have a significant capacity to lower triglyceride and cholesterol levels. Fish oil has been shown to significantly elevate HDL. In fact the evidence for fish oil's ability to lower triglyceride levels has been so strong that the FDA has actually approved a prescription brand of fish oil to treat elevated triglyceride levels in humans. Although approved as a prescription drug, many studies have shown that non-prescription fish oil can also lower triglyceride levels, as well.

Many physicians have turned to fish oil and other natural remedies to further reduce cholesterol levels beyond that of statins. In a 2005 study in Japan, over eighteen thousand people with a history of coronary disease were asked to take 1,800 mg of omega-3, EPA daily together with a statin. Based on a significant additional reduction of cholesterol because of the added EPA fish oil, the study subjects had a nineteen percent additional reduction in risk of major heart events such as heart attacks.

Findings like this have inspired many physicians to use a natural treatment such as fish oil to further reduce cholesterol levels without increasing the dose of the statin. Despite its many benefits, statins can have significant side effects. Statin therapy can cause muscle cramping, and in severe cases, muscle breakdown can occur. Additionally, statins can deplete an important antioxidant in the body called CoQ-10. CoQ-10 is needed for basic cell function, that naturally diminishes as people age, and may be low in people with heart conditions, diabetes, cancer, Parkinson's disease, and other conditions.

Other vitamins and supplements have also been used to lower cholesterol. These include B-Vitamins, red rice yeast, soy and soluble fiber in addition to omega-3 fish oils and plant phytosterols as previously discussed. The bottom-line is, despite the warnings, as a society we are continuing to eat foods that will elevate cholesterol levels. For the foreseeable future statins will continue to be used by healthcare providers to lower cholesterol levels and reduce heart and vascular disease risk factors. The good news is there are also a number of natural alternatives that can be used alone and as added treatment to statins.

If you have any medical condition and are on prescription drugs, including statins, you should ask your healthcare provider before taking any additional dietary supplements. Although safe for most people, some supplements can have unwanted side effects if mixed with certain medications.

How's Your Blood Pressure?

It's estimated that thirty-one percent of Americans are hypertensive, and twenty percent have high blood pressure that has not been diagnosed. Because there are no symptoms of this condition, it is not surprising that many people are unaware that they have hypertension. The lack of symptoms is why hypertension is known as "the silent killer". Undiagnosed or improperly managed hypertension can lead to several serious conditions including aneurysms, coronary artery disease, heart failure, kidney failure, and stroke.

Of people who have sought treatment for hypertension, only forty-seven percent are adequately controlled. There are a number of studies indicating that getting more exercise, cutting back on sodium, and taking fish oil supplements can reduce blood pressure, increase the efficacy of antihypertensive drugs, and decrease the risk of cardiovascular disease. In 2014, P. Miller and colleagues published a very comprehensive analysis of seventy rigorously controlled clinical trials and concluded that the combination of EPA and DHA in fish oil is effective in reducing both systolic and diastolic blood pressure.

Fish Oil and Heart Disease

There are many challenges in comparing the results of one clinical trial to another. This is especially true if the patients or subjects who are being evaluated are not hospitalized or confined to the study area. The vast majority of trials conducted throughout the world are conducted among people who are living at home. Even though different research teams might be studying the exact same treatment, and following identical procedures, their results and findings might vary. One could write volumes on all the variables that might possibly affect the outcome of clinical trials, but, for your benefit, I'll be brief.

The average age of each group is likely to be different. If the studies are in different countries, diets are surely different. People's individual eating habits and patterns within a country will be different. What about their genetic makeup? What about members of different cultures within a country's population? Do participants exercise similarly? Do they all work? How stressful are their jobs? What about vitamins, supplements and other medications people might be taking? Researchers try to adjust for subjects' level of education and other known variables, but the playing field typically turns out to be uneven.

That's the bad news. The good news is that researchers have a reliable way of analyzing data from several trials, even if the trial's methodology or subject's profiles weren't identical. The process, known as meta-analysis, enables researchers to spot patterns of similarity, or other interesting relationships that may come to light in the context of multiple studies.

M. Casula and colleagues published the findings of one such meta-analysis in 2013. The research team looked at reported on eleven studies, which included over fifteen thousand patients. In total, they reviewed three hundred and sixty reports and concluded that their "results provide further evidence on the positive effect of omega-3 fatty acid supplementation administered with dosage greater one gram per day for at least one year in preventing cardiac death, sudden death and myocardial infarction among patients with a history of cardiovascular disease".

CHAPTER

11

Effects of Diet on Your Health

Today we have more food choices than one could begin to imagine just a few decades ago. Supermarkets offer an ever-increasing selection of healthful fruits and vegetables, and along with them not so healthful packaged and prepared foods. Magazines and TV cooking shows constantly tempt us with intriguing dishes. Restaurants featuring food from every corner of the globe saturate our big cities and suburbs, as well. Despite this cornucopia of interesting, exotic, and delicious food, as we a nation we are not eating better. In fact, many would suggest we are eating worse. One might say the Western diet is a recipe for disaster.

The evidence surrounds us. Excessive body weight due to accumulated fat is linked to coronary heart disease, high blood pressure, stroke sleep apnea, Type 2 diabetes, high cholesterol, gallstones, reproductive problems, and osteoarthritis. The National Institute for Health tells us:

- More than two in three adults are considered to be overweight or obese
- More than one in three adults are considered to be obese
- More than one in twenty adults are considered to have extreme obesity
- About one-third of children and adolescents ages six to nineteen are considered to be overweight or obese
- More than one in six children and adolescents ages six to nineteen are considered to be obese

It Hasn't Always Been this Way

Our hunter-gatherer ancestors' limited meat supply came from low-fat wild game as opposed to higher-fat domesticated animals. A larger proportion of their diet came from fruits and nuts. Wheat and rice were not yet being farmed, so carbohydrate intake was much lower. Also, hunting and gathering meant a lot of walking— perhaps fifteen kilometers a day—and a lot of work. In his book, *"The Story of the Human Body: Evolution, Health and Disease"*, Daniel Lieberman refers to forensic evidence that our Stone Age ancestors were free of many diet-related diseases that plague us today, such as tooth decay and type 2 diabetes.

In 1822, the average American consumed about nine grams of sugar a day. That's about two teaspoons and less than a third of an ounce. Today, according o the U.S. Department of Agriculture, we consume between one hundred and fifty to one hundred and seventy pounds of sugar a year! Per capita use of flour and cereal products was one hundred and thirty-eight pounds in the 1970s. By the turn of this century, average consumption increased about forty-five percent to reach two hundred pounds. It is known that high intake of simple sugars, wheat and other grains, as well as a dietary imbalance between omega-3 EFAs and omega-6 EFAs all contribute inflammation.

How Do We Know When We're Overweight?

An extra five or ten pounds on a typical adult body might make one's clothes a little tight, but that would not be considered overweight. With diet and exercise that problem could be resolved in a month or two; sooner if need be. Overweight and obesity are loosely defined as body weight that is higher than what might be considered healthy for a given height. To identify exactly where people fit on the normal weight to obesity spectrum, we use the Body Mass Index (BMI). This guide, widely used by healthcare professionals, uses a formula based on height and weight to determine what percentage of one's body is fat. The following chart applies to adults. It's a good tool to assess where you are and to establish a goal for weight loss. (The Centers for

Disease Control and Prevention website has a calculator to determine BMI for children and teens.)

Body Mass Index for Adults

Height	21	22	23	24	25	26
4'10"	100	105	110	115	119	124
5'0"	107	112	118	123	128	133
5'1"	111	116	122	127	132	137
5'3"	118	124	130	135	141	146
5'5"	126	132	138	144	150	156
5'7"	134	140	146	153	159	166
5'9"	142	149	155	162	169	176
5'11"	150	157	165	172	179	186
6'1"	159	166	174	182	189	197
6'3"	168	176	184	192	200	208

Once you've found your weight, move to the top of that column to determine your BMI.

What Does Body Mass Index Mean?

BMI	
18.5–24.9	Normal weight
25.0–29.9	Overweight
30.0–39.9	Obese
40.0 and above	Extreme obesity

Summary

Summary of Omega-3 Fish Oil Benefits

There are now thousands or articles and placebo-controlled clinical studies in a variety of inflammatory conditions and diseases demonstrating the human health benefits of omega-3s, both EPA and DHA. Nearly all have shown that supplementation with omega-3s has significant benefits with regard to changes in signs, symptoms, and disease outcomes. Omega-3s have synergistic action with other anti-inflammatory treatments and can often allow lower doses of anti-inflammatory medications. In some cases even eliminate the need for NSAIDs.

Unfortunately Americans don't get enough omega-3s in their diets. A study done by Harvard researchers in 2009 estimated that over eighty percent of Americans are deficient in dietary sources of omega-3. The researchers concluded that over ninety-six thousand people will die annually due to complications (mostly heart related) associated with omega-3 deficiency.

Patients with various mental illnesses, including depression, attention-deficit/hyperactivity disorder, along with age-related dementia and Alzheimer's disease have benefited from higher levels of omega-3. Although the mechanisms of action in these conditions are not fully clear, they are most likely related to the effect of omega-3s modulating the immune system, interacting with the inflammatory response, improved synaptic function and providing the needed cellular membrane building block, DHA, to build new nerve cells. Healthy vision and prevention of degenerative diseases of the eye are also believed to benefit from adequate omega-3 for similar reasons. Perinatal supplementation with omega-3s is critical, and benefits infant brain and retinal development.

Other diseases that have benefited from fish oil supplementation are inflammatory diseases such as macular degeneration, rheumatoid arthritis,

Crohn's disease, ulcerative colitis, psoriasis, migraine headaches, and asthma. Chronic inflammation is considered a contributing factor to atherosclerosis (thickening of blood vessel walls and reduction of blood flow). Numerous studies have shown a positive effect of omega-3s in lowering the incidence of ischemic heart disease and myocardial infarction, as well as the risk of atrial fibrillation.

Omega-3 consumption also lowers plasma triglycerides, resting heart rate, and blood pressure, along with improved heart filling and efficiency, lower inflammation, and improved vascular function. The overall result is a reduced risk of cardiac death for those with higher levels of omega-3. As a result, the American Heart Association recommends intake of about one gram of omega-3 fish oil for prevention of coronary artery disease and two to four grams per day for people with high triglycerides.

Numerous studies investigating rheumatoid arthritis, inflammatory bowel disease and dysmenorrhea, along with the study we published in spine pain patients, demonstrate that omega-3 supplementation can significantly benefit pain control. EPA/DHA supplementation reduces patient-assessed joint pain intensity, morning stiffness, number of painful and/or tender joints, and therefore reduces nonsteroidal anti-inflammatory drug (NSAID) consumption.

A comprehensive review of the role of omega-3 supplementation in inflammatory bowel disease, although outcomes have been variable, report improved gut healing, decreased disease activity, decreased use of steroids, and decreased relapse.

Because most Americans are deficient in this essential fatty acid, in our opinion if you were to choose only one dietary supplement to take daily it would be omega-3 fish oil. Not all fish oil supplements are the same. Knowing how your fish oil supplement is processed, stored and what quality measures and purity is found in each capsule are critical components when buying a fish oil supplement. Ratios of EPA to DHA do matter when you are concerned that your LDL cholesterol could be elevated from too much DHA.

We hope with your new knowledge on the benefits of omega-3 fish oil you develop our same passion to prevent disease before it begins. Taking fish oil is a start, but you need to do more. Start today to make better lifestyle choices by increasing your exercise, eating the right foods, avoiding toxins and tobacco smoke, and controlling stress. With these four pillars you can make the right choice for better health.

References

Acheson A, Conover JC, Fandl JP, DeChiara TM, Russell M, Thadani A, Squinto S P, Yancopoulos GD, Lindsay RM. A BDNF autocrine loop in adult sensory neurons prevents cell death. *Nature.* March 1995;374(6521):450-3.

Albert CM, et al. Blood levels of longchain n-3 fatty acidsand the risk of sudden death. N Engl J Med 2002; 346: 1113–18. GISSI-Prevenzione Investigators. Dietary supplementation with n-3 polyunsaturated fatty acids and vitamin E after myocardial infarction: Lancet 1999; 354: 447–55.

Ammann EM, et al. Omega-3 fatty acids and domain-specific cognitive aging: secondary analyses of data from WHISCA. *Neurology.* October 2013;81(17):1484-91.

AREDS2 Research Group, Effect of Omega-3 Fatty Acids, Lutein/Zeaxanthin, or Other Nutrient Supplementation on Cognitive Function: The AREDS2 Randomized Clinical Trial. JAMA. 2015 Aug 25;314(8):791-801

AREDS2 Research Group, Effect of long-chain ω-3 fatty acids and lutein + zeaxanthin supplements on cardiovascular outcomes: results of the Age-Related Eye Disease Study 2 (AREDS2) randomized clinical trial. JAMA Intern Med. 2014 May;174(5):763-71.

Bang, H.O., Dyerberg, J., Nielsen, A.B., Plasma lipid and lipoprotein pattern in Greenlandic west-coast Eskimos, Lancet 297 (7710) (1971) 1143e1146.

Bauer I, Hughes M, Rowsell R, Cockerell R, Pipingas A, Crewther S, Crewther D. Omega-3 supplementation improves cognition and modifies brain activation in young adults. *Hum Psychopharmacol.* March 2014;29(2):133-44.

Bays, H, Clinical Overview of Omacor: A Concentrated Formulation of Omega-3 Polyunsaturated Fatty Acids, Am J Cardiol 2006;98[suppl]:71i–76i

Berson EL, et al. Further evaluation of docosahexaenoic acid in patients with retinitis pigmentosa receiving vitamin A treatment: subgroup analyses. *Arch Ophthalmol*. September 2004;122(9):1306-14.

Beydoun MA, Fanelli Kuczmarski MT, Beydoun HA, Hibbeln JR, Evans MK, Zonderman AB. Omega-3 fatty acid intakes are inversely related to elevated depressive symptoms among United States women. *J Nutr*. November 2013;143(11):1743-52.

Bhargava, R, Kumar, P., A Randomized Controlled Trial of Omega-3 fatty Acids in Dry Eye Syndrome, Int J Ophthalmol, Vol 6, No. 6 Dec 18 2013

Brasky TM, et al. Plasma phospholipid fatty acids and prostate cancer risk in the SELECT trial. *J Natl Cancer Inst*. August 2013;105(15):1132-41.

Brasky, Theodore M., Serum Phospholipid Fatty Acids and Prostate Cancer Risk: Results From the Prostate Cancer Prevention Trial, American Journal of Epidemiology Advance Access published April 24, 2011

Brenda L. Plassman, Bo Qin, et al., Fish Intake Is Associated with Slower Cognitive Decline in Chinese Older Adults, J. Nutr. 2014

Carter JR, Schwartz CE, Yang H, Joyner MJ. Fish oil and neurovascular reactivity to mental stress in humans. *Am J Physiol Requl Integr Comp Physiol*. April 2013;304(7):R523-30.

Casula M, Soranna D, Catapano AL, Corrao G. Long-term effect of high dose omega-3 fatty acid supplementation for secondary prevention of cardiovascular outcomes: A meta-analysis of randomized, placebo controlled trials. *Atheroscler Suppl*. August 2013;14(2):243-51.

Celecoxib capsule labeling by G.D. Searle LLC Division of Pfizer, *Inchttp://labeling.pfizer.com/ShowLabeling.aspx?id=793*, July 26, 2016

Center for Disease Control and Prevention, "High Cholesterol Facts", *http://www.cdc.gov/cholesterol/facts.htm* Cited July 26, 2016

Chavarro JE, Stampfer MJ, Hall MN, Sesso HD, Ma J. A 22-y prospective study of fish intake in relation to prostate cancer incidence and mortality. *Am J Clin Nutr.* 2008;88:1297-303.

Chowdhury R, et al. Association between fish consumption, long chain omega 3 fatty acids, and risk of cerebrovascular disease: systematic review and meta-analysis. *BMJ.* October 2012;345:e6698.

Colcombe SJ, Erickson KI, Scalf PE, et al. Aerobic exercise training increases brain volume in aging humans. *Gerontol A Biol Sci Med Sci.* 2006;61(11):1166-1170.

Coletta JM, Bell SJ, Roman AS. Omega-3 fatty acids and pregnancy. *Rev Obstet Gynecol.* 2010;3(4):163-171.

Consumer Reports News, "New dietary guidelines call on societal changes for our better health June 17, 2010", http://www.consumerreports.org/cro/news/2010/06/new-dietary-guidelines-call-on-societal-changes-for-our-better-health/index.htm Cited July 26, 2016.

Consumer Reports News, "Choose the Right Fish To Lower Mercury Risk Exposure: The government wants you to eat more seafood. The key to limiting your risk is choosing the right fish August 2014", http://www.consumerreports.org/cro/magazine/2014/10/can-eating-the-wrong-fish-put-you-at-higher-risk-for-mercury-exposure/index.htm, Cited July 26, 2016

Daiello LA, Gongvatana A, Dunsiger S, Cohen RA, Ott BR. Association of fish oil supplement use with preservation of brain volume and cognitive function. *Alzheimers Dement.* February 2015;11(2):226-35.

Danaei G, Ding EL, Mozaffarian D, Taylor B, Rehm J, Murray CJ, Ezzati M. The preventable causes of death in the United States: comparative risk assessment of dietary, lifestyle, and metabolic risk factors. *PLoS Med.* April 2009;6(4):e1000058.

Daviglus ML, et al. Fish consumption and the 30-year risk of fatal myocardial infarction. N Engl J Med 1997; 336: 1046–53. Albert CM, et al. Fish consumption and risk of sudden cardiac death. JAMA 1998; 279: 23–28.

DeDea L. When to take statins; Lovaza versus OTC fish oil supplements. *JAAPA*. May 2011;24(5):23.

Dickinson A, Boyon N, Shao A: Physicians and nurses use and recommend dietary supplements: report of a survey. Nutr J. 2009, 8: 29-10.1186/1475-2891-8-29.

Farooqui AA, Ong WY, Horrocks LA, Chen P, Farooqui T. Comparison of biochemical effects of statins and fish oil in brain: The battle of the titans. *Brain Res Rev*. December 2007;56(2):443-71.

Ferris LT, Williams JS, Shen CL. The effect of acute exercise on serum brain-derived neurotrophic factor levels and cognition function. *Med Sci Sports Exerc*. 2007;39(4):728-734.

Endo, Akira, A historical perspective on the discovery of statins, Proc Jpn Acad Ser B Phys Biol Sci. 2010 May 11; 86(5): 484–493.

Fontani G, Corradeschi F, Felici A, Alfatti F, Migliorini S, Lodi L. Cognitive and physiological effects of Omega-3 polyunsaturated fatty acid supplementation in healthy subjects. *Eur J Clin Invest*. 2005;35(11):691-9.

Food and Drug Administration, "FDA Cuts Trans Fatin Processed Foods". *http://www.fda.gov/downloads/ForConsumers/ConsumerUpdates/UCM451467. pdf*, Cited July 26, 2016.

Food and Drug Administration, "FAERS Reporting by Patient Outcomes by Year". *http://www.fda.gov/Drugs/GuidanceComplianceRegulatoryInformation/ Surveillance/AdverseDrugEffects/ucm070461.htm*. Cited July 26, 2016.

Food and Drug Administration, "FDA Strengthens Warning of Heart Attack and Stroke Risk for Non-Steroidal AntiInflammatory Drugs", *http://www.*

fda.gov/downloads/ForConsumers/ConsumerUpdates/UCM454066.pdf Cited July 26, 2016.

Food and Drug Administration, "Medication Guide for Non-Steroidal Anti-Inflammatory Drugs," *(NSAID) http://www.fda.gov/downloads/Drugs/DrugSafety/ucm089162.pdf* Cited July 26, 2016.

Food and Drug Administration, « Fish: " What Pregnant Women and Parents Should Know", *http://www.fda.gov/Food/FoodborneIllnessContaminants/Metals/ucm393070.htm,*Cited July 26, 2016

Freund Levi Y, et al. Transfer of omega-3 fatty acids across the blood-brain barrier after dietary supplementation with a docosahexaenoic acid-rich omega-3 fatty acid preparation in patients with Alzheimer's disease: the OmegAD study. *J Intern Med.* April 2014;275(4):428-36.

Gani OA. Are fish oil omega-3 long-chain fatty acids and their derivatives peroxisome proliferator-activated receptor agonists? *Cardiovasc Diabetol.* March 2008;7:6.

Ginty, AT., Conklin, SM., .Short-termsupplementationofacutelong-chainomega-3 polyunsaturated fatty acids may alter depression status and decrease symptomology among young adults with depression: A preliminary randomized and placebo controlled trial, PsychiatryResearch229(2015)485–489

Greenberg, JA., Bell, SJ., Omega-3 Fatty Acid Supplementation During Pregnancy, Rev Obstet Gynecol. 2008 Fall; 1(4): 162–169.

GSK Source, "Lovaza® Highlights of Prescribing", *https://www.gsksource.com/lovaza* Cited July 26, 2016

Hanbauer I, et al. The decrease of n-3 fatty acid energy percentage in an equicaloric diet fed to B6C3Fe mice for three generations elicits obesity. *Cardiovasc Psychiatry Neurol.* 2009;2009:867041.

Hansen-Krone IJ, et al. High fish plus fish oil intake is associated with slighty reduced risk of venous thromboembolism: the Tromsø Study. *J Nutr.* June 2014;144(6):861-7.

He K, Xun P, Brasky TM, Gammon MD, Stevens J, White E. Types of fish consumed and fish preparation methods in relation to pancreatic cancer incidence: the VITAL Cohort Study. *Am J Epidemiol.* January 2013;177(2):152-60.

Heller AR, Fischer S, Rössel T, Geiger S, Siegert G, Ragaller M, Zimmermann T, Koch T. Impact of n-3 fatty acid supplemented parenteral nutrition on haemostasis patterns after major abdominal surgery. *Br J Nutr.* January 2002;87 Suppl 1:S95-101.

Hibbeln, J R., Fish consumption and major depression, The Lancet, Volume 351, Issue 9110, 18 April 1998, Pages 1213

Hibbeln, J R., Seafood consumption, the DHA content of mothers' milk and prevalence rates of postpartum depression: a cross-national, ecological analysis. J Affect Disord. 2002 May;69(1-3):15-29.

Holly LT, Blaskiewicz D, Wu A, Feng C, Ying Z, Gomez-Pinilla F. Dietary therapy to promote neuroprotection in chronic spinal cord injury. *J Neurosurg Spine.* August 2012;17(2):134-40.

Horrobin DF. The membrane phospholipid hypothesis as a biochemical basis for the neurodevelopmental concept of schizophrenia. Schizophr Res. 1998;30:193–208.

Horrocks LA, Farooqui AA. Docosahexaenoic acid in the diet: its importance in maintenance and restoration of neural membrane function. Prostaglandins Leukot Essent Fatty Acids. 2004;70:361–72.

Hu, F.B., et al., Television watching and other sedentary behaviors in relation to risk of obesity and type 2 diabetes mellitus in women. JAMA, 2003. 289(14): p. 1785-91

Inoue, S., et al., Television viewing time is associated with overweight/obesity among older adults, independent of meeting physical activity and health guidelines. J Epidemiol, 2012. 22(1): p. 50-6.

Innis, SM., Dietary (n-3) Fatty Acids andBrainDevelopment, 2007 American Society for Nutrition. J. Nutr. 137: 855–859, 2007.

International Food Information Council Foundation, "2012 Food & Health Survey", *http://www.foodinsight.org/2012 Food Health Survey Consumer Attitudes toward Food Safety Nutrition and Health* Cited July 26, 2016

Jacka FN, Pasco JA, Williams LJ, Meyer BJ, Digger R, Berk M. Dietary intake of fish and PUFA, and clinical depressive and anxiety disorders in women. *Br J Nutr.* June 2013;109(11):2059-66.

Jenkins DJ, Kendall CW, Marchie A, et al. Direct comparison of a dietary portfolio of cholesterol-lowering foods with a statin in hypercholesterolemic participants. Am J Clin Nutr. 2005;81:380–7.

Johnston DT, Deuster PA, Harris WS, Macrae H, Dretsch MN. Red blood cell omega-3 fatty acid levels and neurocognitive performance in deployed U.S. Servicemembers. *Nutr Neurosci.* January 2013;16(1):30-8.

Lanza, FL., Chan, FKL, Prevention of NSAID-Related Ulcer Complications, Am J Gastroenterol 2009; 104:728–738; 2009.115;

Kepler CK, Huang RC, Meredith D, Kim JH, Sharma AK. Omega-3 and fish oil supplements do not cause increased bleeding during spinal decompression surgery. *J Spinal Disord Tech.* May 2012;25(3):129-32.

Kiecolt-Glaser JK, et al. Omega-3 fatty acids, oxidative stress, and leukocyte telomere length : A randomized controlled trial. *Brain Behav Immun.* February 2013;28:16-24.

Koski RR. Omega-3-acid ethyl esters (Lovaza) for severe hypertriglyceridemia. *P T.* May 2008;33(5):271-303.

Kremer JM, et al. Effects of high-dose fish oil on rheumatoid arthritis after stopping nonsteroidal anti-inflammatory drugs. Clinical and immune correlates. *Arthritis Rheum.* August 1995;38(8):1107-14.

Kris-Etherton PM, Harris WS, Appel LJ. Fish consumption, fish oil, omega-3 fatty acids, and cardiovascular disease. *Circulation.* November 2002;106(21):2747-57.

Kris-Etherton PM, Harris WS, Appel LJ. Omega-3 fatty acids and cardiovascular disease: new recommendations from the American Heart Association. *Arteriosclerosis Thrombosis and Vascular Biology.* February 2003;23(2):151-2.

Land, W. Diets Could Prevent Many Diseases, Lipids 38(4):317-21 · April 2003

Lassek, W.D., Gaulin, S.J.C. Waist-hip ratio and cognitive ability: is gluteofemoral fat a privileged store of neurodevelopmental resources? Evolution and Human Behavior 29 (2008) 26–34

Lassek WD, Gaulin SJ. Sex differences in the relationship of dietary Fatty acids to cognitive measures in American children. *Front Evol Neurosci.* November 2011;3:5.

Lau CS, Morley KD, Belch JJ. Effects of fish oil supplementation on non-steroidal anti-inflammatory drug requirement in patients with mild rheumatoid arthritis—a double-blind placebo controlled study. *Br J Rheumatol.* November 1993;32(11):982-9.

Laurin D, et al. Physical activity and risk of cognitive impairment and dementia in elderly persons. *Arch Neurol.* 2001;58:498-504.

Lev-Tzion R, Griffiths AM, Ledder O, Turner D, Omega 3 fatty acids (fish oil) for maintenance of remission in, Crohn's disease (Review) The Cochrane Library, 2014, Issue 2

Leventakou V, et al. Fish intake during pregnancy, fetal growth, and gestational length in 19 European birth cohort studies. *Am J Clin Nutr.* March 2014;99(3):506-16.

Lewis MD, Bailes J. Neuroprotection for the warrior: dietary supplementation with omega-3 fatty acids. *Mil Med.* October 2011;176(10):1120-7.

Lewis MD, Ghassemi P, Hibbeln J. Therapeutic use of omega-3 fatty acids in severe head trauma. *Am J Emerg Med.* January 2013;31(1):273.e5-8.

Lieberman, D., The Story of the Human Body: Evolution, Health, and Disease, Pantheon Books, 2013

Life Extension Magazine, "VAP Cholesterol Testing 2007", *http://www. lifeextension.com/magazine/2007/5/report_vap/page-01*, Cited July 26, 2016

Ma QL, Teter B, Ubeda OJ, Morihara T, Dhoot D, Nyby MD, Tuck ML, Frautschy SA, Cole GM. Omega-3 fatty acid docosahexaenoic acid increases SorLA/LR11, a sorting protein with reduced expression in sporadic Alzheimer's disease (AD): relevance to AD prevention. *J Neurosci.* December 2007;27(52):14299-307.

Macpherson H, Silberstein R, Pipingas A. Neurocognitive effects of multivitamin supplementation on the steady state visually evoked potential (SSVEP) measure of brain activity in elderly women. *Physiol Behav.* 2012;107(3):346-354.

Maroon JC, The Longevity Factor: How Resveratrol and Red Wine Activate Genes for a longer and Healthier Life, Simon and Schuster 2009.

Maroon JC, Bost JW, Borden MK, Lorenz KM, Ross NA. Natural anti-inflammatory agents for pain relief in athletes. *Neurosurg Focus.* 2006;21(4)

Maroon JC, Bost JW. Fish Oil: The Natural Anti-Inflammatory. *Basic Health Publications.* 2006.

Maroon JC, Bost JW. Omega-3 Fatty acids (fish oil) as an anti-inflammatory: an alternative to nonsteroidal anti-inflammatory drugs for discogenic pain. *Surgical Neurology.* April 2006;65:326-331.

Miller PE, Van Elswyk M, Alexander DD. Long-chain omega-3 fatty acids eicosapentaenoic acids and docosahexaenoic acid and blood pressure: a meta-analysis of randomized controlled trials. *Am J Hypertens.* July 2014;27(7):885-96.

Mills JD, Hadley K, Bailes JE. Dietary supplementation with the omega-3 fatty acid docosahexaenoic acid in traumatic brain injury. *Neurosurgery.* February 2011;68(2):474-81.

Mozaffarian D, Wu JH. Omega-3 fatty acids and cardiovascular disease: effects on risk factors, molecular pathways, and clinical events. *J Am Coll Cardiol.* November 2011;58(20):2047-67.

National Institute of Health, "Cognitive training shows staying power, NIH-funded trial shows 10-year benefit in realms of reasoning, speed", https://www.nia.nih.gov/newsroom/2014/01/cognitive-training-shows-staying-power, Cited July 26, 2016

Narendran R, Frankle WG, Mason NS, Muldoon MF, Moghaddam B. Improved working memory but no effect on striatal vesicular monoamine transporter type 2 after omega-3 polyunsaturated fatty acid supplementation. *PLoS One.* 2012;7(10):e46832

Nikolakopoulou Z, Nteliopoulos G, Michale-Titus AT, Parkinson EK. Omega-3 polyunsaturated fatty acids selectively inhibit growth in neoplastic oral keratinocytes by differentially activating ERK1/2. *Carcinogenesis.* December 2013;34(12):2716-25.

Norris JM, et al. Omega-3 polyunsaturated fatty acid intake and islet autoimmunity in children at increased risk for type 1 diabetes. *JAMA.* September 2007;298(12):1420-8.

O'Callaghan N, Parletta N, Milte CM, Benassi-Evans B, Fenech M, Howe PR. Telomere shortening in elderly individuals with mild cognitive impairment may be attenuated with omega-3 fatty acid supplementation: a randomized controlled pilot study. *Nutrition*. April 2014;30(4):489-91.

Orengo IF, Black HS, Wolf JE, Jr. Influence of fish oil supplementation on the minimal erythema dose in humans. Arch Dermatol Res. 1992;284:219-221.

Oken E, Belfort MB. Fish, fish oil, and pregnancy. *JAMA*. October 2010;304(15):1717-8.

Patel JV, Tracey I, Hughes EA, Lip GY. Omega-3 polyunsaturated fatty acids: a necessity for a comprehensive secondary prevention strategy. *Vasc Health Risk Manag*. 2009;5:801-10.

Pottala JV, Yaffe K, Robinson JG, Espeland MA, Wallace R, Harris WS, Higher RBC EPA + DHA corresponds with larger total brain and hippocampal volumes: WHIMS-MRI Study. (2014) Neurology. 82(5): 435-42 and Epub 2014 Jan 22

Qin B, Plassman BL, Edwards LJ, Popkin BM, Adair LS, Mendez MA. Fish intake is associated with slower cognitive decline in Chinese older adults. *J Nutr*. October 2014;144(10):1579-85.

Rhodes LE, O'Farrell S, Jackson MJ, Friedmann PS. Dietary fish-oil supplementation in humans reduces UVB-erythemal sensitivity but increases epidermal lipid peroxidation. J Invest Dermatol. 1994;103:151-1

Rhodes LE, Shahbakhti H, Azurdia RM, et al. Effect of eicosapentaenoic acid, an omega-3 polyunsaturated fatty acid, on UVR-related cancer risk in humans. An assessment of early genotoxic markers. Carcinogenesis. 2003;24:919-925.

Rizos EC, Ntzani EE. [Omega]-3 fatty acids and lutein + zeaxanthin supplementation for the prevention of cardiovascular disease. *JAMA Intern Med*. 2014;174(5):771-772.

Rizos EC, Ntzani EE, Bika E, Kostapanos MS, Elisaf MS. Association between omega-3 fatty acid supplementation and risk of major cardiovascular disease events: a systematic review and meta-analysis. *JAMA*. September 2012;308(10):1024-33.

Sands SA, Reid KJ, Windsor SL, Harris WS. The impact of age, body mass index, and fish intake on the EPA and DHA content of human erythrocytes. *Lipids*. April 2005;40(4):343-7.

Saunders EFH, Reider A, Singh G, Gelenberg AJ, Rapoport SI. Low unesterified:esterified eicosapentaenoic acid (EPA) plasma concentration ratio is associated with bipolar disorder episodes, and omega-3 plasma concentrations are altered by treatment. Bipolar Disord 2015: 17: 729–742. 2015

Sears B, Bailes J, Asselin B. Therapeutic uses of high-dose omega-3 fatty acids to treat comatose patients with severe brain injury. *PharmaNutrition*. July 2013;1(3):86-89.

Sears B, Ricordi C. Anti-inflammatory nutrition as a pharmacological approach to treat obesity. *J Obes*. 2011;2011.

Simopoulos, A.P. The importance of the omega-6/omega-3 Fatty Acid ratio in cardiovascular disease and other chronic diseases. Exp. Biol. Med. 2008, 233, 674–688

Simopoulos AP. Evolutionary aspects of diet, the omega-6/omega-3 ratio and genetic variation: nutritional implications for chronic diseases. *Biomed Pharmacother*. November 2006;60(9):502-7.

Simopoulos AP. Omega-3 fatty acids in health and disease and in growth and development. *Am J Clin Nutr*. Spetember 1991;54(3):438-63.

Singleton RH, Yan HQ, Fellows-Mayle W, Dixon CE. Resveratrol attenuates behavioral impairments and reduces cortical and hippocampal loss in a rat controlled cortical Healthcare model of traumatic brain injury. *J Neurotrauma*. June 2010;27(6):1091-9.

Sirot V, Leblanc JC, Margaritis I. A risk-benefit analysis approach to seafood intake to determine optimal consumption. *Br J Nutr.* June 2012;107(12):1812-22.

Spite M, et al. Resolvin D2 is a potent regulator of leukocytes and controls microbial sepsis. *Nature.* October 2009;461(7268):1287-91.

Stoll AL, et al. Omega 3 fatty acids in bipolar disorder: a preliminary double-blind, placebo-controlled trial. *Arch Gen Psychiatry.* May 1999;56(5):407-12.

Tanskanen A, Hibbeln JR, et al., Fish consumption and depressive symptoms in the general population in Finland. Psychiatr Serv. 2001 Apr;52(4):529-31

US Department of Agriculture, "Dietary Guidelines for Americans 2010", https://health.gov/dietaryguidelines/dga2010/dietaryguidelines2010.pdf, Cited July 26, 2016

US Department of Agriculture, Agriculture Fact Book Chapter 2 - Profiling Food Consumption in America, http://www.usda.gov/factbook/chapter2.pdf, Cited July 26, 2016

Vakhapova V, Cohen T, Richter Y, Herzog Y, Kam Y, Korczyn AD. Phosphatidylserine containing omega-3 fatty acids may improve memory abilities in nondemented elderly individuals with memory complaints: results from an open-label extension study. *Dement Geriatr Cogn Disord.* 2014;38(1-2):39-45.

Virtanen JK, Mursu J, Voutilainen S, Uusitupa M, Tuomainen TP. Serum omega-3 polyunsaturated fatty acids and risk of incident type 2 diabetes in men: the Kuopio Ischemic Heart Disease Risk Factor study. *Diabetes Care.* 2014;37(1):189-96.

Wallace A, Chinn D, Rubin G. Taking simvastatin in the morning compared with in the evening: randomised controlled trial. *BMJ.* 2003;327(7418):788

Wang J, et al. Omega-3 polyunsaturated fatty acids enhance cerebral angiogenesis and provide long-term protection after stroke. *Neurobiol Dis.* August 2014;68:91-103.

Whalley LJ, Fox HC, Wahle KW, Starr JM, Deary IJ. Cognitive aging, childhood intelligence, and the use of food supplements: possible involvement of n-3 fatty acids. *Am J Clin Nutr.* December 2004;80(6):1650-7.

Wu A, Ying Z, Gomez-Pinilla F. Dietary omega-3 fatty acids normalize BDNF levels, reduce oxidative damage, and counteract learning disability after traumatic brain injury in rats. *J Neurotrauma.* October 2004;21(10):1457-67.

Wu A, Ying Z, Gomez-Pinilla F. Omega-3 fatty acids supplementation restores mechanisms that maintain brain homeostasis in traumatic brain injury. *J Neurotrauma.* October 2007;24(10):1587-95.

Xin W, Wei W, Li XY. Short-term effects of fish-oil supplementation on heart rate variability in humans: a meta-analysis of randomized controlled trials. *Am J Clin Nutr.* May 2013;97(5):926-35.

Yaffe K, Barnes D, Nevit M, et al. A prospective study of physical activity and cognitive decline in elderly women. *Arch Intern Med.* 2001;161:1703-1708.

Yates A, Maroon JC, Bost J. et al. Evaluation of Lipid Profiles and the Use of Omega-3 Essential Fatty Acid in Professional Football players. *Sports Health.* Jan 2009;1(1):21-30

Yokoyama M. Effects of eicosapentaenoic acid (EPA) on major cardiovascular events in hypercholesterolemic patients: the Japan EPA Lipid Intervention Study (JELIS). American Heart Association Scientific Sessions 2005; November 13-16, 2005; Dallas, Texas. Late Breaking Clinical Trials II.

Yokoyama M, Origasa H; JELIS Investigators. Effects of eicosapentaenoic acid on cardiovascular events in Japanese patients with hypercholesterolemia: rationale, design, and baseline characteristics of the Japan EPA Lipid Intervention Study (JELIS). Am Heart J. 2003;146:613-620.

Yon MA, Mauger SL, Pickavance LC. Relationships between dietary macronutrients and adult neurogenesis in the regulation of energy metabolism. *Br J Nutr.* May 2013;109(9):1573-89.

Zheng JS, Hu XJ, Zhao YM, Yang J, Li D. Intake of fish and marine n-3 polyunsaturated fatty acids and risk of breast cancer: meta-analysis of data from 21 independent prospective cohort studies. *BMJ.* June 2013;346:f3706.